World Sustainability Series

Series Editor

Walter Leal Filho, European School of Sustainability Science and Research, Research and Transfer Centre "Sustainable Development and Climate Change Management", Hamburg University of Applied Sciences, Hamburg, Germany

Due to its scope and nature, sustainable development is a matter which is very interdisciplinary, and draws from knowledge and inputs from the social sciences and environmental sciences on the one hand, but also from physical sciences and arts on the other. As such, there is a perceived need to foster integrative approaches, whereby the combination of inputs from various fields may contribute to a better understanding of what sustainability is, and means to people. But despite the need for and the relevance of integrative approaches towards sustainable development, there is a paucity of literature which address matters related to sustainability in an integrated way.

Notes on the quality assurance and peer review of this publication

Prior to publication, the works published in this book are initially assessed and reviewed by an in-house editor. If suitable for publication, manuscripts are sent for further review, which includes a combined effort by the editorial board and appointed subject experts, who provide independent peer-review. The feedback obtained in this way was communicated to authors, and with manuscripts checked upon return before finally accepted. The peer-reviewed nature of the books in the "World Sustainability Series" means that contributions to them have, over many years, been officially accepted for tenure and promotion purposes.

Hasnan Baber · Mina Fanea-Ivanovici
Editors

Sustainable Financing—A Contemporary Guide for Green Finance, Crowdfunding and Digital Currencies

Sustainable Financing

Springer

Editors
Hasnan Baber
Department of Management
Strategy and Entrepreneurship
School of Business Administration
American University of Sharjah
Sharjah, United Arab Emirates

Mina Fanea-Ivanovici
Department of Economics and Economic
Policies
Bucharest University of Economic Studies
Bucharest, Romania

ISSN 2199-7373 ISSN 2199-7381 (electronic)
World Sustainability Series
ISBN 978-3-031-80971-2 ISBN 978-3-031-80969-9 (eBook)
https://doi.org/10.1007/978-3-031-80969-9

This Springer imprint is published by the registered company Springer Nature Switzerland AG
The registered company address is: Gewerbestrasse 11, 6330 Cham, Switzerland

Acknowledgments

We wish to express our heartfelt appreciation to our esteemed reviewers whose invaluable contributions have been essential in the development of this book. Their expertise and feedback have played a significant role in shaping the content and maintaining its quality. This book would not have been possible without their dedicated efforts, and we are sincerely thankful for their unwavering commitment to excellence.

Iqra Farooq
Ewelina Idziak
Rashmita Singh
Rashid Khalil
Domenica Federico
Muhammad Azhar Khalil
Néstor Soler del Toro
Hristina Mikić
Ali Akbar Firoozi
Faizan Ashraf Mir
Peterson K. Ozili
Saurabh Sood

Hasnan Baber
Mina Fanea-Ivanovici

Contents

Contributors

Rosa Adamo Department of Business and Law, University of Calabria, Arcavacata of Rende, CS, Italy

Hasnan Baber Department of Management, Strategy and Entrepreneurship, School of Business Administration, American University of Sharjah, Sharjah, United Arab Emirates

Itishree Behera KIIT School of Management (KSOM), KIIT University, Bhubaneswar, Odisha, India

Nestor Soler del Toro Malaga, Spain

Mina Fanea-Ivanovici Department of Economics and Economic Policies, Bucharest University of Economic Studies, Bucharest, Romania

Domenica Federico Department of Social and Human Sciences, eCampus University, Novedrate, CO, Italy

Ali Akbar Firoozi Department of Civil Engineering, Faculty of Engineering and Technology, University of Botswana, Gaborone, Botswana

Ali Asghar Firoozi Department of Civil Engineering, Faculty of Engineering, National University of Malaysia (UKM), selangor, Malaysia

Shubhangi Vitthal Gaikwad School of Economics and Commerce, MIT World Peace University, Pune, India

Samreedhi Gupta Jindal Global Law School, O.P. Jindal Global University, Sonipat, Haryana, India

Arif Hasan ABS, Amity University, GwaliorMadhya Pradesh, India

Bincy Baburaj Kaluvilla The Emirates Academy of Hospitality Management, Dubai, United Arab Emirates

Muhammad Azhar Khalil Department of Finance and Risk Management, MSME Business School, Assumption University, Bangkok, Thailand

Rashid Khalil Department of Banking and Finance, School of Business, Bahrain Polytechnic University, Isa Town, Bahrain

Vandana Kushwaha School of Economics and Commerce, MIT World Peace University, Pune, India

Ali Asghar Firoozi Department of Civil Engineering, Faculty of Engineering, National University of Malaysia (UKM), selangor, Malaysia

Hristina Mikić Faculty of Management, Metropolitan University, Belgrade, Serbia

Faizan Ashraf Mir Central University of Kashmir, Tulmulla, J&K, India

Pragyan Nanda Parul University, Vadodara, Gujarat, India

Antonella Notte Department of Social and Human Sciences, eCampus University, Novedrate, CO, Italy

Peterson K. Ozili Central Bank of Nigeria, Abuja, Nigeria

Pathathai Sinliamthong Department of Finance and Risk Management, MSME Business School, Assumption University, Bangkok, Thailand

Saurabh Sood Jindal Global Law School, O.P. Jindal Global University, Sonipat, Haryana, India

Daria Elisa Vuc Doctoral School of Economics I, Bucharest University of Economic Studies, Bucharest, Romania

Archana Yadav GLA University Mathura, Mathura, UP, India

List of Figures

List of Tables

Introduction to Sustainable Finance: Green Finance, Crowdfunding and Digital Currencies

Hasnan Baber and Mina Fanea-Ivanovici

Abstract In today's world, the concept of sustainable finance is gaining momentum as individuals and businesses alike become more aware of the impact of their financial decisions on the environment and society.

1 Introduction

In today's world, the concept of sustainable finance is gaining momentum as individuals and businesses alike become more aware of the impact of their financial decisions on the environment and society. With the rise of climate change and social inequality, there is an increasing need for financial institutions to adopt sustainable practices that support long-term environmental and social goals. This book will provide a contemporary guide to green finance crowdfunding and digital currencies highlighting the key principles and benefits of sustainable finance in the modern economy.

Sustainable finance involves integrating environmental social and governance (ESG) factors into financial decisions playing a crucial role in driving investment toward sustainable development and the Sustainable Development Goals (SDGs) (Capital Markets and Sustainable Finance 2023). It has evolved from being a niche concern to a mainstream practice influencing business mindsets and investment strategies (Edmans and Kacperczyk 2022). It involves investing in projects and companies that promote sustainability and ethical practices while avoiding those that harm the environment or society. The concept emphasizes non-predatory finance that

H. Baber (✉)
Department of Management, Strategy and Entrepreneurship, School of Business Administration, American University of Sharjah, PO Box 26666, Sharjah, United Arab Emirates
e-mail: hbaber@aus.edu

M. Fanea-Ivanovici
Department of Economics and Economic Policies, Bucharest University of Economic Studies, 6 Piata Romana, District 1, 010374 Bucharest, Romania
e-mail: mina.ivanovici@economie.ase.ro

H. Baber and M. Fanea-Ivanovici (eds.), *Sustainable Financing—A Contemporary Guide for Green Finance, Crowdfunding and Digital Currencies*, World Sustainability Series, https://doi.org/10.1007/978-3-031-80969-9_1

generates value while fostering sustainable development through positive externalities and investments in green and social initiatives (Goglio and Catturani 2019). Financial institutions are key players in advancing sustainable development yet this role is often overlooked by financial professionals and policymakers (Jeucken 2010). Sustainable finance not only contributes to financial objectives but also addresses broader societal and environmental goals making it a critical component of modern finance.

This book on Sustainable Finance offers a comprehensive and contemporary guide to green finance crowdfunding blockchain and digital currencies providing a professional insight into the evolving landscape of sustainable finance. Through sections like 'The Green Shift: How Alternative Financing is Going Sustainable' and 'Exploring Green Bonds as a Catalyst for Climate Financing in India' readers delve into the innovative ways finance is being used to drive environmental impact. A systematic review of sustainable finance coupled with discussions on blockchain and digital assets equips readers with a deep understanding of the mechanisms shaping the industry. From exploring the nuances of crowdfunding including its applications in medical crowdfunding and political engagement to elucidating how crowdfunding is transforming environmental and social initiatives this book presents a holistic view of the potential for financial tools to drive positive change. Delving into crowdfunding insights from NGOs to private ventures sheds light on the power of collective action for impactful causes making it a must-read for anyone looking to navigate the world of finance with a sustainability lens and embrace the potential of digital currencies for a better future.

2 Green Finance

Green finance refers to the financing of environmentally friendly investment policies and projects aimed at promoting sustainability and combating climate change. It encompasses funding for green investments in areas like environmental goods and services climate damage prevention and financial mechanisms specifically designed for green initiatives (Lindenberg 2014; Berensmann and Lindenberg 2019). Green finance plays a crucial role in the global shift towards sustainability by involving various financial asset classes and institutions such as green banking debt markets and structured funds driven by institutional investors regulatory bodies and central banks (Jaiwant and Kureethara 2023). This financial approach is essential for addressing environmental challenges reducing carbon footprints and promoting eco-friendly business practices contributing to a sustainable financial system that balances economic growth with environmental preservation (Martin 2023; Al-Sheryani and Nobanee 2020).

Green finance promotes sustainable financing for environmental projects fostering economic growth while respecting environmental and social aspects. It reduces climate change risks preserves macroeconomic stability and develops new financing sources (Martin 2023). Green finance aids in funding eco-friendly projects combating climate change achieving SDGs fulfilling NDC obligations promoting green growth and mobilizing resources through initiatives like sovereign green bonds (Ravichandran and Roy 2022). Green finance offers reduced carbon emissions supports economic growth and enhances environmental sustainability through renewable energy projects. It attracts funding from both the public and private sectors for environmental preservation (Selvapandian et al. 2022). The role and significance of green finance in creating a sustainable financial system to address the issues of environmental degradation are vital and similarly, the role and importance of green money in creating and developing eco-friendly systems that will decrease ozone-layer depletion control pollution and increase green activities are prevalent (Jaiwant and Kureethara 2023).

3 Crowdfunding

Crowdfunding has emerged as a powerful tool in sustainable finance for social entrepreneurs providing them with a platform to access funding from a broad base of individuals who believe in their mission and impact (Baber and Fanea-Ivanovici 2022). This innovative approach has also revolutionized the way medical crowdfunding operates allowing individuals facing high medical costs to raise funds from a supportive network of contributors (Snyder et al. 2016). Moreover, crowdfunding has facilitated political engagement by enabling candidates and organizations to connect directly with their supporters and raise funds for their causes (Fanea Ivanovici and Baber 2023). Beyond that crowdfunding has played a crucial role in mobilizing resources for environmental and social good projects empowering communities to take action on issues such as climate change conservation and social justice (Prędkiewicz and Kalinowska-Beszczyńska 2020). Overall crowdfunding continues to shape the landscape of financing offering a democratized and efficient way to support a wide range of initiatives that drive positive change in society (Fanea-Ivanovici and Baber 2021).

4 Digital Currencies

Digital currencies particularly central bank digital currencies (CBDCs) and blockchain technology play a significant role in promoting sustainable finance and environmental sustainability (Yang et al. 2023; Arnone 2022; Paradise 2022). CBDCs like the digital RMB have been shown to increase the issuing of green bonds reduce emissions and improve green land ratios thus positively impacting sustainable development (Shen 2022). Blockchain technology on the other hand has been effective in enhancing green production tracking environmental data and facilitating real-time decision-making for eco-friendly practices (Ezeji and Ezekiel 2022). Additionally the FinTech revolution in China driven by developments in CBDCs and green digital finance is gradually becoming a crucial tool in Chinese foreign policy contributing to sustainable development in countries participating in initiatives like the Belt and Road Initiative. These advancements highlight the potential of digital currencies in fostering sustainable finance and environmental responsibility (Paradise 2022).

Digital currencies such as cryptocurrencies and blockchain technology are also playing a growing role in sustainable finance. Blockchain technology in particular offers a secure and transparent way to track and verify transactions making it a valuable tool for promoting trust and accountability in the financial sector. Cryptocurrencies on the other hand provide an alternative means of exchange that can facilitate borderless and efficient transactions reducing the need for traditional banking systems and their associated environmental footprint.

In conclusion sustainable finance is a vital component of our transition to a more sustainable and equitable future. By incorporating ESG criteria into financial decision-making processes investing in green finance initiatives leveraging crowdfunding for sustainability and embracing digital currencies we can drive positive change and create a more resilient economy. As individuals and businesses we have the power to make a difference through our financial choices—let's make them count for the betterment of our planet and society.

Meta-description: Learn about sustainable finance and how green finance crowdfunding and digital currencies are changing the financial landscape for a more sustainable future.

References

Al-Sheryani K, Nobanee H (2020) Green finance: a mini-review. SSRN Electron J. https://doi.org/10.2139/ssrn.3538696

Arnone G (2022) Blockchain and cryptocurrency innovation for a sustainable financial system. Int J Industr Manage 15(1):1–16. https://doi.org/10.15282/ijim.15.1.2022.8994

Baber H, Fanea-Ivanovici M (2022) Fifteen years of crowdfunding—a bibliometric analysis. Technol Anal Strat Manage: 1–15. https://doi.org/10.1080/09537325.2022.2089548

Berensmann K, Lindenberg N (2019) Green finance: across the universe. In: Corporate social responsibility ethics and sustainable prosperity, pp 305–332

Capital Markets & Sustainable Finance (2023) World investment report 2023, 97–138. https://doi.org/10.18356/9789210027847c009

Edmans A, Kacperczyk M (2022) Sustainable finance. Rev Fin 26(6):1309–1313. https://doi.org/10.1093/rof/rfac069

Ezeji MO, Ezekiel AO (2022) Digital finance the future and sustainability of the Nigeria banking system: a review. Int J Res Fin Manage 5(2):205–211. https://doi.org/10.33545/26175754.2022.v5.i2c.169

Fanea Ivanovici M, Baber H (2023) Using the civic voluntarism model to compare the donation intentions in US and India political crowdfunding. Int J Electron Govern 15(2). https://doi.org/10.1504/ijeg.2023.10057043

Fanea-Ivanovici M, Baber H (2021) Crowdfunding model for financing movies and web series. Int J Innov Stud 5(2):99–105. https://doi.org/10.1016/j.ijis.2021.06.001

Goglio S, Catturani I (2019) Sustainable finance: a common ground for the future in Europe? Palgrave Stud Impact Fin: 239–261. https://doi.org/10.1007/978-3-030-22510-0_10

Jaiwant SV, Kureethara JV (2023) Green finance and fintech: toward a more sustainable financial system. Green Fin Instruments FinTech Investment Strat: 283–300. https://doi.org/10.1007/978-3-031-29031-2_12

Jeucken M (2010) Sustainable finance and banking. Routledge. http://books.google.ie/books?id=pDNXsbIL-k8Candprintsec=frontcoveranddq=Sustainable+finance+and+banking:+The+financial+sector+and+the+future+of+the+planetandhl=andcd=1andsource=gbs_api

Lindenberg N (2014) Definition of green finance. German Development Institute, p 3

Martin V (2023) Green finance: regulation and instruments1. J Central Banking Theor Pract 12(2):185–209. https://doi.org/10.2478/jcbtp-2023-0019

Paradise JF (2022) The role of green digital finance in achieving sustainable development goals in china's belt and road initiative. Econ Law Inst Asia Pacific: 167–185. https://doi.org/10.1007/978-981-19-2662-4_8

Prędkiewicz K, Kalinowska-Beszczyńska O (2020) Financing eco-projects: analysis of factors influencing the success of crowdfunding campaigns. Int J Entrep Behav Res 27(2):547–566. https://doi.org/10.1108/ijebr-05-2020-0339

Ravichandran S, Roy M (2022). Green finance: a key to fight with climate change. Ind J Econ Fin 2(2):34–38. https://doi.org/10.54105/ijef.b2526.112222

Selvapandian G, Jeyapaul PP, Gunabalan B (2022) Adoption of green financing strategies with renewable energy resources for global economic growth. Glob Econ J 22(04). https://doi.org/10.1142/S2194565923500045

Shen C (2022) Digital RMB, RMB internationalization and sustainable development of the international monetary system. Sustainability 14(10):6228. https://doi.org/10.3390/su14106228

Snyder J, Mathers A, Crooks VA (2016) Fund my treatment!: a call for ethics-focused social science research into the use of crowdfunding for medical care. Soc Sci Med 169:27–30. https://doi.org/10.1016/j.socscimed.2016.09.024

Yang Q, Zheng M, Wang Y (2023) The role of CBDC in green finance and sustainable development. Emerg Mark Financ Trade 59(15):4158–4173. https://doi.org/10.1080/1540496x.2023.2185096

The Green Shift: How Sustainable Financing is Going Sustainable

Bincy Baburaj Kaluvilla

Abstract Green finance has become a driving power that revolutionizes the world's financial landscape by directing investments in eco-friendly projects and initiatives. While having common goals with sustainable finance, like fostering innovation and sustainability, green finance differs in its emphasis on environmental issues together with the use of specialized financial instruments. This chapter addresses the notion of green finance as a special branch of sustainable finance, focusing on its part in promoting a pollution-free environment and sustainable development. Green bonds and loans represent a novel way of funding environmentally friendly projects, providing feasible financial solutions for startups and businesses that aim to infuse sustainability into everyday operations. This chapter develops a critical discussion of these instruments, emphasizing their expansion, consequences, and propensity to become mainstream. The chapter starts by characterizing green bonds and loans, particular instances of their prosperous issuances, and data regarding these markets' size and geographic distribution. It goes on to analyze the influence of these tools on corporate sustainability and innovation regarding academic studies and cases. A consolidated discussion of the regulatory landscape framing green finance is given, involving a comparative examination of main regulations and taxonomies across regions, such as the EU Taxonomy and the Green Bond Principles. The chapter presents benchmarks and areas of concern within the regulatory system. The possibility of digital innovation, particularly blockchain technology, to improve transparency and accountability in green finance is discussed, dealing with the issue of greenwashing. Examples of how the principles of green finance may be applied in practice are offered via an extended case study of green bond issuances conducted by Amazon, with detailed specifics of the financed projects and their green and financial results. The chapter highlights the author's original analysis and insights, noting trends, challenges, and opportunities within green finance and suggesting policy measures for governing agents and investors. The focus of this section on green bonds and loans offers a detailed, actionable analysis of these innovative funding

B. B. Kaluvilla (✉)
The Emirates Academy of Hospitality Management, Dubai, United Arab Emirates
e-mail: baburaj_bincy@yahoo.co.in

H. Baber and M. Fanea-Ivanovici (eds.), *Sustainable Financing—A Contemporary Guide for Green Finance, Crowdfunding and Digital Currencies*, World Sustainability Series, https://doi.org/10.1007/978-3-031-80969-9_2

channels, making a substantial contribution to the wider discussions of sustainable finance and its role in achieving positive environmental change.

Keywords Green finance · Sustainable finance · Sustainability · Green bonds · Green loans · Environmental regulation · Blockchain · Amazon

1 Introduction

This chapter covers green finance as a multi-dimensional concept through rigorous discussions on its theoretical foundations, the particular tools it uses, a feedback loop between regulation and technology, pertinent obstacles to accountants, and insights into the future. Green finance has a solid foundation because it is rooted in the idea that long-term economic progress depends on short-term ecological sustainability. The ESG concept, which concentrates on the part of the environmental, social, and governance elements for investment decision-making, has recently received a lot of interest. As a result of this phenomenon, the index of global ESG assets under management reached up to $35.3 trillion in 2020 (Díaz-Peña et al. 2022). As compensation for its issuing, the green bond provides an opportunity to finance a project aimed at a certain environmental purpose by a contributing investor. The share of green bonds in the global market has significantly increased as the issuance in 2020 reached a renewed peak of $269.5 billion from the Climate Bonds Initiative (CBI) (2021a, b). Governments and financial regulators are key factors in establishing and performing green finance by creating the environment, policies, and regulations of the market and the integrity. The coexistence of technology and regulation in green finance also develops the chances and challenges. Fintech can make the green financing process more successful by increasing transparency, decreasing costs, and extending the financing access network. Even though laws governing and regulating technology are prevalent, data security, privacy, and fraud remain crucial.

By utilizing a thorough literature review and universally acceptable data statistics and cases, this chapter shows a comprehensive understanding of green finance as an sustainable finance subsector. Amazon's case study also showcases a huge corporation's practical concept of green finance that can leverage various green finance instruments to support its sustainable objectives and gradually employ the low-carbon economy. The chapter aims to elaborate on the trades and issues in the green financing space to increase the knowledge of sustainable finance globally and its role in attaining the desired environmental remedy. Thus, the role of green finance is greatly illuminated in resolving ecological problems and in the transformation to a new environment that is sustainable and resilient. This subsequently emphasizes the necessity of formulating favorable regulatory tools and seeing the bright prospects that the technology provides for the sector.

2 Background

As a change factor in the financial world, green finance has contributed toward allocating funds to green and ecological investments. Shift happens due to increased awareness of the world-changing magnitude and urgency of climate change and many other important environmental issues. As seen in the Global Sustainable Investment Alliance data, the global sustainable investment assets of 2020 more than doubled to trillions from 15% in 2018, which shows a shift in investors' decision-making in the environmental direction. Sustainable finance is similarly goal-oriented to sustainable financing, which involves driving innovation and great sustainability. The Green Finance market segregates itself attentively to environmental issues and wide use of specialized financial instruments (Fang and Shao 2022; Liu and Wu 2023). This chapter deals with the green finance topic, a functional branch of specialized sustainable finance as a tool for overcoming environmental problems (Gilchrist et al. 2021; Fu and Irfan 2022; Bhutta et al. 2022). The main reason for green finance is to solve urgent environmental problems, including global warming and resource exhaustion, by changing the course of capital investments while concentrating on projects that positively impact the environment (Taghizadeh-Hesary and Yoshino 2019). Nowadays, climate change is one of the major elements that green finance deals with. As per the Paris Agreement, the global goal is to keep the average increase in global temperatures below 2 °C, which is already higher than before the industrial revolution period (Gao et al., 2017). This goal of the International Energy Agency projects that by 2030, these clean energy investments will reach $4 billion annually (Raimi et al., 2022). Green finance is an impetus for more funding to materialize toward this shift, and it also supports the United Nations' Sustainable Development Goals (SDGs) (Barua 2019).

3 Conceptualizing Green Finance and Sustainable Financing

Green finance needs to be explained and linked to sustainable finance before the details of green finance are addressed. Green finance may be defined as financial investments and instruments that target environmental sustainability and fund projects with positive environmental outcomes (Liu and Wu 2023). The main categories of such projects are renewable energy, energy efficiency, sustainable transportation, and green buildings (Gilchrist et al. 2021). The development of green finance was aimed at addressing age environmental issues related to global warming, earth-planet pollution, and resource deformation by diverting capital flow and navigation into sustainable projects and activities. Capital flows of this kind are critical for the shift to a low-carbon and resource-utilized economy as they finance the development and deployment of clean technology and sustainable infrastructure. Sustainable financing uses various financial instruments, such as green bonds, loans with sustainability targets, green equity investments, and sustainable insurance. In addition, the Devices serve the purpose of stabilizing economic activities on the green, thus covering the monetary risks related to environmental degradation and climate change. The development of green finance is due to many reasons like the increasing understanding of the seriousness of environmental issues, the development of new financial technologies and business models, and supportive policy and regulatory environments emerging."

Meanwhile, sustainable financing is the broader array of financial channels and methods outside conventional banking systems (Carè et al. 2018). This involves crowdfunding, peer-to-peer lending, invoice financing, and venture capital. Reformation of finance has become an answer to restrictions and barriers of classic banks, which fail to meet financing of small and medium enterprises, startups, and left-out groups. Digital technologies and innovative business models enable sustainable finance platforms to create direct links between borrowers and investors, avoiding intermediaries and transaction costs. Disintermediation allows sustainable finance to reach more customers and provide customized financing solutions. Although sustainable finance and green finance aim to stimulate innovation and sustainability, green finance stands out by focusing on environmental issues and using specific financial instruments (Ozili 2022). On the other hand, sustainable finance is focused mainly on increased access to finance and financial inclusion with little or no regard for the environmental application of the funded projects.

"In this sense, green finance becomes one of the specialized categories of sustainable finance, applying creative financial instruments to propagate environmental sustainability (Neumann 2023). The novelty of green finance instruments like green bonds and loans places them as substitutes for traditional financial instruments, and reiterate green finance as part of the sustainable finance domain (Liu and Wu 2023). For instance, green bonds are fixed-income securities that fund environmentally sustainable investments, such as renewable energy infrastructure, energy-efficient buildings, and clean transportation. Another term is green loans, which are financing mechanisms for green projects and are usually tied to the environmental performance of the borrower. These instruments are distinct from traditional bonds and loans because they are purposefully created to promote environmentally sustainable activities, and they must comply with defined green criteria and reporting obligations. The growing tendency of green finance as sustainable finance refers to the rising need to unify the financial markets with environmental objectives and to attract private funds to the sustainable economy. Through targeted financial instruments and investments into environmentally friendly projects, green finance can become one of the key forces of response to rapidly emanating environmental issues that the earth is currently dealing with."

4 The Evolution of Green Finance: Trends and Impacts

As the attention to environmental issues went through an upsurge and several countries began focusing on climate change problems, the green finance sector has been on the rise, rapidly developing in recent years. The scientific research by Dervi et al. (2022) conducted a bibliometric analysis covering a 4-decade period and identified primary sources, journals, and countries that have the most impact on the green and socially responsible financing discussion. This makes the picture clearer, and the idea of how green finance will change the behavior of investors and performance models becomes visible. The facts confirm that green finance research accelerates almost ten times faster than in the 2000s, mainly sustainable investments, investors' disclosure of green issues, and the economic status of the mentioned investments. Bibliometric analysis further shows the interdisciplinary angle of the green finance study, as evidenced by the fact that it is anchored within economics, finance, environmental science, and social science. The authors have annotated these organs, mentioning some of the journals that are in the waste sector, like the Journal of Cleaner Production, Sustainability, and Sustainable Finance and Investment. Besides, a table gives information about the most prolific research countries in green finance. Here, the US, China, and the UK are listed first. The paper presents findings on the evolution process of sustainable finance and its growing influence on academics and practitioners.

Green financing instruments are growing fast as conscious consumerism increases the environmental standards for goods and services in agricultural, industrial, and other sectors (Khan et al. 2022). Consumers nowadays, when making purchasing decisions, strongly consider their impact on the environment. The conscious consumers who aim to go green by choosing green products thus emerge. It has also meant that businesses must be more green-conscious and invest in new technologies and infrastructure that will enable them to do their part. For that reason, being a source of funding for the corporate sector's transition to sustainability is one of the new roles for environmental financing. For instance, green finance can help this sector fund the adoption of farming practices such as incline farming, precision farming, and agroforestry. The finance sector is frequently applied for the industry's clean-tech planning and deployment (such as renewable energy, energy efficiency, and circular economy actions). These developments are thus pointing towards how green finance can help steer these sectors toward environment- and nature-friendly practices. Green finance is anticipated to grow sizeably as companies and investors become interested in sustainability's long-term financial and branding returns.

Carried mainly by private banks, green financial instruments have become highly popular in East Asia–Pacific countries (Liu et al. 2023; Taghizadeh-Hesary and Yoshino 2019; Lalon 2015). Among the new actors in the green finance market over the past few years are China, Japan, and South Korea, of which the successors are driven by market demand, government policies, and international climate commitments. For instance, in the case of China, the government uses different green finance techniques, such as establishing a national green development fund, issuing green bonds, and introducing green credit guidelines for the banks. As a result, China has become the green bonds market leader in the world, with a total of $44.3 billion in green bonds issued last year alone (Climate Bonds Initiative 2021). In Japan and South Korea, private bankers, on their part, are already assisting this by introducing green loans for loans and developing numerous more green investment products. These nations also undertake global green finance, for example, through the programs offered by the "United Nations Environment Program Finance Initiatives and Sustainable Banking Network." Green finance in East Asia–Pacific is growing due to the expanding recognition of the link between sustainable investment and a low-carbon economy, as well as the importance of financial institutions in their contribution to sustainable development.

"Green finance is an instrument offering visual evidence of the impact of this financial instrument on environmental sustainability. The Global Label of Climate Bonds Initiative showed that the emission of green bonds amounted to $269.5 billion in 2020, more than the previous year by 1.1% during the tough period brought by the COVID-19 pandemic. The size of this growth demonstrates the ability of green finance to shape and establish itself in sustainable development. Sustainable financial instruments, specifically green bonds, represent the flow of money directed into projects that improve the environment, like renewable energy, energy efficiency, and clean transport. The finance capital raised by the issues of green bonds is allocated to green projects, and it is subject to strict reporting and auditing standards to ensure that it is used for its original goal of nature. The green bond market has been thriving in the last few years because, in the fast-developing capitalistic view, investors crave more investment tools to support environmental sustainability. Besides, the financial advantages of environmentally responsible projects are being considered nowadays. Furthermore, other green finance instruments, including green and sustainability-linked loans, have grown rapidly in recent years. The Loan Market Association assessment (2021) shows that green loans and sustainability-linked issuance worldwide reached $197bn in 2020, an increase of 58% over 2019. They not only help finance the green projects but also reward successful borrowers for reaching and either cutting down on greenhouse gas emissions or increasing the amount of renewable energy."

Green financing has also crucially impacted the investing sanctions of investors and financial institutions. With increasing demand for green investments, investors now address ESG criteria while making investor decisions (Schumacher 2023). Investment policies that are sustainable or stable aim at a lot of things, including the rising financial risks that are caused by climate change, investor demands that the issue of sustainability be considered, and the discovery these investments result in financial growth in the long run. The report by "Global Sustainable Investment Alliance (2021) reveals that global sustainable investment" assets grew to USD 35.3 trillion in 2020, which constitutes over 15% growth compared to 2018 (Global Sustainable Investment Review, 2021). These figures demonstrate the progress in the journey towards incorporating sustainable investing and the materiality of ESG issues in financials. Besides ESG factors, which have also recently led to new financial products and services, such as ESG-themed funds, green bonds, and sustainability-linked loans, being introduced to the market. These products allow investors to customize their investment choices in line with their values and help divert the mainstream economy into a more ecologically friendly sector. The sharp increase in sustainable investing has brought a positive change in the stipulation of ESG performance and disclosure from companies since investors are now focusing on the importance of environmental matters and social issues.

5 Green Bonds and Loans: Instruments for Sustainable Development

"Green bonds and loans, a primary source of investment in projects to mitigate climate change, have expanded rapidly. Most green bonds are issued to finance projects with environmentally friendly projections, including constructing environmentally friendly energy sources, energy efficiency, and sustainable transport infrastructure (Zhao et al., 2022). These bonds are issued across a widely varied range of issuers, with the bond proceeds dedicated towards green projects described in the bond prospectus and set by guidelines in the prospectus. According to the Climate Bonds Initiative (2021), a milestone of green bonds reached a record of over $269 billion in 2020, and the incoming years were characterized by the rapid growth of this market segment. This rapid expansion results from the growing interest of investors in taking into account sustainable manufacturing and green bonds as instruments to finance the transformation toward a low-carbon economy. Besides the nature of the international bond market, the top three countries in green bond issuances in 2020 have been the US (with $50.1 billion), Germany (with $40.2 billion) and France ($32.1 billion) (Climate Bonds Initiative 2021)."

A green loan signifies using a loan instrument for supporting environment-oriented projects, and the target amount of this instrument tends to be lower than that of the green bond (Dursunöz et al. 2023). The most common type of such loans is provided by banks or other financial institutions to companies or projects that set up themselves adhering to certain environmental thresholds such as greenhouse gas mitigation or sustainable land use. Green loans can be used to invest in projects such as renewable energy installation, making efficiency better, and developing sustainable infrastructure. Though it was relatively small, the green loan market also showed outstanding growth. It was only $59 billion in 2018 but rose to $80 billion in 2019 (BloombergNEF 2020). The continuously growing green loan lending and the evolving perspectives among economy decision-makers about green finance as a tool for developing a clean economy and other entities' growing demand for sustainable funding motivate market growth. Green finance gives businesses a wide range of opportunities and chances to have a more versatile approach to investments in their sustainability projects, which is commensurate with each project.

Researchers explored the link between green bonds and loans to corporations' social responsibility and progressive entrepreneurship. These supportive instruments include a focus on achieving environmental targets as well as corporations' profit and image. Goal-oriented activities by Gilchrist et al. (2021) have been introduced to ensure the realization of both corporate goals and a better environment. A study examining the green bond issuance from US corporations from 2013 to 2018 on a sample basis found that the announcement of green bond issuance is accompanied by a surge in abnormal stock returns and higher ESG ratings (Environmental, social, and governance) of the company. In the authors' opinion, green bonds may signify a company's focus on environmental care and thus lead to a rise in public esteem and the influx of socially conscious investors. Moreover, they get the additional facts that companies with green bonds as their source of financing are spending more on their research and development and filing more patents concerning new renewable technologies. Though green bonds typically carry higher interest rates, investors tend to be more willing to take the risk due to their solid financial background and the environmental goals they can achieve.

Meanwhile, Wang et al. (2022) once again emphasized the benefits that green bonds provide to projects that lack the finance and the one that creates a green innovational ecosystem. In the research of a sample of Chinese firms that have listed their green bonds in the market of 2016–2020, the authors saw a significant growth of green patent applications and green innovation outputs as soon as the green bond was released. The authors make it clear that green bonds would enable a wide variety of investors to reach their goals at lower financial costs, resulting in fewer financial constraints and better prospects for investing in green innovation. In addition, the research also revealed that green bonds had a larger influence on environmental innovation in more highly environmentalist-regulated industries and companies with better corporate governance structures. As the outcome of the above case proves, green bonds provide the best approach to be applied, particularly in the strict environmental regulations and corporate governance mechanisms. Consequently, 'green bonds' are a vital means for authorities and investors who are involved in the search for sustainable development, as well as for the creation of a decarbonized economy.

"Furthermore, the standard and the certification bodies have emerged in the context of green bonds and loans market development to ensure the authenticity of these products. Formulated by the International Capital Market Association (ICMA), the Green Bond Principles (GBP) is a set of voluntary guidelines for issuing green bonds, which, in a way, provide investors with information on how the issuer determines environmental benefits (ICMA 2024). The GBP outlines four core components that define a green bond: beneficiary identification, proceeds used effectively and efficiently, project selection and evaluation, implementation of project action plans, and reporting. They aim to ensure that transactions in the green bonds market are typically free of commercial, political, and sector bias. In contrast, the proceeds of these bonds are spent on real environmental issues. Similar to the aforementioned Green Loan Principles (GLP) introduced by the Loan Market Association (LMA), it acts as a standard for green loans. GLP basis also has the same four main components of GBP and is intended to be inherent in the unanimity and uniformity of green loans."

"The standards and certification schemes have significantly influenced the growth of the green bond and loan market as they improve investors' confidence in the environmental quality of such instruments. Nonetheless, green bond and loan credibility and effectiveness issues are unresolved. An additional problem is that there is no standard definition for the term green project, which causes worries about greenwashing and misuse of funds. Another challenge is the need for more sophisticated and standardized impact reporting frameworks that allow investors to evaluate the environmental and societal outcomes of green bond and loan investments. However, expanding the green bond and loan market represents a major opportunity to channel capital to achieve sustainable development goals and facilitate the transition towards a low-carbon and sustainable economy. In this respect, policymakers, investors, and issuers have the critical function to maintain the stable development of this market in the future."

6 The Regulatory Landscape and Technological Intersection in Green Finance

The regulatory environment is a crucial part of developing and implementing green finance. Andreeva et al. (2018) stress the necessity of new theoretical and methodological approaches to create an efficient financial framework for the green economy because ecological, economic, social, and geopolitical risks continuously redefine the global financial landscape. The authors propose a transversal trend that brings social, ecological, and economic dimensional convergence since the existing regulating system cannot tackle the multi-dimensional challenges of green economy transformation. They offer a holistic model that includes fiscal incentives, regulatory norms, and market-based tools as instruments for green finance "targeting." Framework also defends international cooperation and joint action, making possible free access for all, and in doing this, it prevents regulatory arbitrage.

The obstacle to "green" finance is that the 'mother tongues' and 'criteria' are not coherent and universal (Berrou et al. 2019). Due to that lack of clarity, the problem of greenwashing can occur, where businesses report things that positively affect the environment through their activities or investments of a green kind. Still, there is no contribution. The problem tackles countries and international organizations by developing green finance frameworks and taxonomies. Such taxonomies should be constructed to define a common lexicon and the scientific principles for notating green economic undertakings as investments. Of course, taxonomies are no simple matter since boundaries are usually set up between stakeholders with different interests and priorities. For example, some sectors may broaden the scope of green activities to include their practice. In contrast, green groups may strongly demand stricter regulations or standards to ensure genuine environmental benefit. The alignment and mandate of conflicting interests is the key to the efficiency and reliability of green finance taxonomies.

"One particular instance is the EU Taxonomy of the European Union, a categorization mechanism for environmentally sustainable economic actions (European Commission 2021). The taxonomy aims to draw a clear and transparent map for investors, companies, and regulators that ensures that the assets classified as sustainable have definite environmental standards. Among sectors that produce 80% of direct carbon dioxide emissions by 2021, energy, manufacturing, transport, and buildings will be put forward in the European Parliament after 2021 (Amanatidis 2021). EU Taxonomy represents a deciding turn in the EU-wide green finance definition unification process and the creation of a limpid framework for investors and companies. It is underpinned by six environmental objectives: Such intersectional concerns should be addressed in the strategy, among others, as climate change adaption, climate change mitigation, sustainable use of water and marine resources, circular economy transition, pollution prevention and control, and biodiversity and ecosystems protection. Within the EU Taxonomy, a socio-economic action is considered environmentally sustainable if it substantially contributes to any given goal and does not seriously hurt any other."

"In addition to the aforementioned, the panel admits the creation of TCFD (Task Force on Climate-related Financial Disclosure) by the Financial Stability Board with a task to develop and promote the disclosure of financial climate information (TCFD 2023). The total sign-ups to TCFD recommendations reached 2600 global organizations by October 2021, with a market cap of about $25 trillion (TCFD 2021). The TRUST of these resources manifests the influence of openness and disclosure on green finance. The TCFD recommendations provide a playing field for the firms to cover the financial information in terms of governance, strategy, risk management, metrics, and targets, which explains how and to what extent climate-related issues and opportunities affect their finances. In the same vein, TCFD allows investors, lenders, and insurers to evaluate climate change risks and the value of opportunities, allowing them to standardize data. To achieve a wide acceptance of TCFD recommendations for businesses and investors, a gradual process will be needed, wherein an understanding of the financial materiality of climate change and the importance of a coherent approach to climate information reporting will develop."

As China's economic and environmental policy stands at a crossroads, tech-finance-legal regulations have received wide attention among researchers and companies. Hu et al.'s approach to regulatory effectiveness for sustainable environmental finance is localized to facilitate adequate digital finance conceptions (Hu et al. 2022). This synergy demonstrates the role of government policies in promoting digital finance and adjusting the environmental regulations that situate society to green technology innovation. The authors posit that digital finance, including online lending platforms and mobile payment systems, can effectively address the financing challenges faced by green technology companies, especially small and medium enterprises (SMEs). By allowing more investors to participate, such investments enhance the financing needs of green technology companies used in developing and commercializing their innovations, which are met by increasing ease of financing. Nonetheless, the authors argue that the impact of digital finance on the greening of technology innovation is conditioned by the quality of environmental regulation. It is found that of all market-based environmental regulations, green taxes and emissions trading schemes are most effective in causing green technology innovation compared to command-and-control regulations like emission standards and technology mandates.

"Han et al. (2023) note that digital finance positively affects the efficiency of green development, while the impact of environmental regulation is not yet positive. Their discovery affirms that digital finance could, in a more significant way, alleviate the financial issues posed by strict environmental regulations of the nature of policy integration. The authors utilize a panel data set of 283 Chinese cities from 2011 to 2019 to analyze the influence of digital finance and environmental regulation on green development efficiency, where the latter is quantified as the city's ability to produce economic output while minimizing environmental pollution and resource consumption. They also note that digital finance contributes significantly to green development efficiency because it facilitates easy access to the finance of green projects and technologies. However, they also note that the effect of environmental regulation on green development efficiency is mixed, with some regulations being positive and some being negative. This may result in regulations that many companies cannot afford, hence resulting in high non-compliance levels."

Fang and Shao's study (2022) represents the role of green finance and its conditions under different environmental regulations. The policy impact of market incentives for environmental conservation and green finance accelerates the innovations related to regional green technologies, but to the contrary, command and control regulation hinders it. This illustrates that the necessary measure is flexible environmental regulations, which would create a way to utilize green finances. Each author utilized 30 Chinese provinces and studied the data from 2009 to 2019 to understand how environmental regulation affects green technology innovation. They also examined how green finance acts as one of the moderators in this process. The decision-makers thus concluded that the carbon market incentive policies, including the recycling of carbon trading schemes and green taxes, can help promote green technology innovation. Instead, they curb the use of command-and-control regulations and embrace emission standards and technology mandates, which are unproductive. Besides, green finance significantly attenuated this earliness, driving favorable impacts and lessening the adverse impacts of market incentive rules. The authors claim that green finance allows corporations to invest in clean technologies and projects, even under super-tight regulatory measures.

7 Challenges and Opportunities in Green Finance: Providing Greenwashing and Exploiting of Blockchain

The development of green finance was preceded by greenwashing, which, in scholars' view, is the most significant one. As Zhang (2023) states, greenwashing is deliberately put forward to make a product, service, or investment appear as if it has positive environmental impacts on the environment, even when it does not. Arguably, it may also lead to the relaxation of the trustworthiness and the worthiness of green finance initiatives and bring about distrust among investors and other stakeholders. There are three main forms of greenwashing: over-talking about the positive environmental features of the product, incomplete disclosure concerning the ecological impact of company practices, and the use of abstract, unsubstantiated definitions of sustainability. Greenwashing constitutes one of the green finance sector's most pressing issues since it contributes to the reduction in the trust size in sustainable investing and the volume of funds ascribed to genuinely environmental projects. This additional chance for greenwashing, in which firms declare environmental effects they do not have to the investors, must be a part of the problem. Furthermore, adequate regulation agencies and checking mechanisms will be emphasized to make the sweet taste of green finance programs linger.

What comes with it is the extent to which the quality and credibility of ESG reporting receive more attention (Dursun-de et al. 2023). A powerful and responsible corporate ESG reporting should be the top-most priority in case investors have to make well-informed decisions and green finance is to realize its potential in driving sustainability. The disclosure of ESG issues is a source of information on the environment, human behavior, and the company it contributes to for investors to assess their portfolio sustainability and long-run viability. However, the standardization and consistency of ESG reporting remains an investigator's challenge because of comparing other companies' sustainability claims and reports. In coincidence, there have been efforts to formulate standard reporting frameworks and bodies like the "Global Reporting Initiative (GRI) and the Sustainability Accounting Standards Board (SASB)." These steps are directed to improve the quality and comparability of the ESG disclosures and to provide the investor with the spotlight on the key details.

Blockchain technology is among the top innovations that can increase transparency and credibility in green finance. Dorofeev (2020) emphasizes that conventional financial institutions would undergo an overhaul due to the blockchain, creating a green bond market and more certainty in providing funds and compliance with a decentralized yet immutable ledger. This is especially true for green finance, where ensuring that those ties will be followed and the funds flow only to the eco-projects is crucial. Blockchain technology is used by green bond issuances and the use of funds to display investors' will; therefore, they can track funds to prove the environmental impact of their investment. Blockchain, a tamper-proof and auditable ledger, can be utilized to prevent greenwashing and trace how funds flow to support sustainable projects in line with their original plans. Moreover, using smart contracts based on the blockchain also allows the confirmation and reporting tasks to be performed, which eventually reduces the cost of compliance and lets environmental performance be automatically monitored in real-time.

By focusing on the case of China, Zhang (2023) discusses how blockchain technology can address the bottlenecks that green finance development faces. The study bases its conclusion on the fact that the innate features of blockchain in the form of distributed ledgers and smart contracts can greatly assist in sharing information, cross-risk elimination, and diversity of green financial products. As a Chinese green finance champion, green financing continues to become a leading development with the government behind this by implementing multiple policies and programs for sustainable investment. Nevertheless, the green finance market in the country might be hindering the adoption of specific standards and the risk of greenwashing. Zhuang realizes that blockchain, as a reliable and transparent technology, can serve as a way to fight against these vulnerabilities by creating a safe green finance network. For instance, blockchain technology can ensure the real-time monitoring of environmental performance through reports and partially verify the process, diminishing the chances of fraud and making green projects more credible worldwide. Besides developing new green financial products, including peer-to-peer lending and green insurance, blockchain has become a fast and trustworthy transaction platform.

8 Amazon's Green Finance Initiatives: A Case Study

"Amazon, a giant multinational corporation, has been successful in embedding green finance into its sustainability strategy. The company aims to reach net-zero carbon emissions by 2040 for its operations and has introduced diverse initiatives to support this goal (Amazon 2022a, b). This commitment goal aligns with the Paris Agreement's aim to limit global warming to below 2 °C above pre-industrial levels. It shows that Amazon is determined to deal with climate change. For that purpose, Amazon has created a holistic sustainability strategy that involves such initiatives as investments in renewable energy, sustainable transport, and energy-efficient buildings. The company has also set up a specialized sustainability team responsible for the company's environmental projects and coordinating with stakeholders on sustainability matters. One of the vital parts of this strategy is Amazon's green finance initiatives, as they provide enough money for the company to invest in sustainability and develop green economy innovations."

"One of the prominent initiatives is the Climate Pledge Fund—a $2 billion venture capital fund set up by Amazon in 2020 to support enterprises producing sustainable technologies and services (Pagitsas, 2022). The fund has invested in multiple companies, primarily in carbon removal, electric aviation, and battery storage advancements, reflecting Amazon's dedication to accelerating green economy innovation. The Climate Pledge Fund is a very large corporate venture capital fund focused on sustainability and shows Amazon's appreciation of the crucial role of innovation in combatting climate change. Amazon supports early-stage companies that produce state-of-the-art sustainable technologies, accelerating the transition to a low-carbon economy and emerging new market possibilities for sustainable products and services. The fund's investments cut across many sectors, such as transport, energy, and agriculture, thus bringing out the multi-dimensional nature of sustainability issues and the need for a comprehensive approach to tackle them."

"Two years before, in May 2021 specifically, Amazon released a $1 billion sustainability bond (Amazon 2021). This bond's net proceeds are applied to finance projects in renewable energy, clean transportation, green buildings, and affordable housing. An objective appraisal by an independent assessor revealed that the bond framework generates favorable environmental and social effects, demonstrating Amazon's commitment to having the financing aligned with sustainability goals (Amazon 2021). Regarding Amazon, the sustainability bond is a powerful achievement since it is the company's first transaction in the green bond market. It shows its willingness to allocate its financial resources to favor sustainable development. The bond follows the Green Bond and Social Bond Principles to guarantee that the funds go into projects benefiting the environment and people and that these benefits are independently verified. Amazon here shows its environmental awareness and the interconnection of the ecological and social issues through the presence of renewable energy, sustainable transportation, environmentally friendly buildings, and cost-effective housing in the bond, which verifies a systematic character of sustainability."

Furthermore, the company has invested vast amounts of money into renewable energy systems, with as many as 310 renewable sources directed towards powering the entirety of its facilities (Amazon 2022a, b), and this allowed the organization to achieve a great rate of about eighty-five percent renewable energy for the day-to-day operations. Its regular electric car fleet, which will hit 100,000 by 2030, proves the firm's commitment to green transport (Gleeson 2023). Renewable energy and clean transport infrastructure are the core elements of Amazon's sustainability strategy, and they also show the company's awareness that it must act fast to transition the world economy to a low-carbon environment. Reinvestment in renewable energy projects helps Amazon decrease its carbon emissions, supports the development of the renewable energy sector, and drives down the cost of clean energy. Likewise, Amazon's investment in electric delivery vehicles also helps promote the development of clean transportation technologies and decreases the environmental effect of logistics operations. Additionally, these investments are important because they help generate job opportunities in the renewable power and clean transportation industries and minimize the risk of the company being affected by fluctuating fossil fuel prices.

Although Amazon's green finance initiatives have been appreciated, the company has also been accused of having a poor environmental record (Rosenblatt 2023). This underlines the difficulties and subtleties of green finance strategies integration in the frames of large companies and the necessity of its improvement and transparency. Though it has sustainability commitments, Amazon has been criticized for its carbon footprint, single-use plastics, and how they affect local communities. These critiques, though, indicate that a holistic approach to sustainability should be employed, where the environmental and social effects of an organization's activities are addressed throughout the value chain. They also highlight the need for transparency and accountability in corporate sustainability reporting, as well as a growing demand from stakeholders to access credible and complete information on companies' environmental and social performance. To meet these challenges, Amazon must invest in green finance projects and collaborate with stakeholders to identify and manage sustainability risks and opportunities. In this respect, ongoing improvement will be necessary, with a readiness to openly and honestly discuss the company's sustainability performance with stakeholders.

9 Conclusion

"The study of green finance as a separate subsector of other financial approaches discloses its dynamic role in empowering the shift to eco-friendly economic models. Environmental considerations are prioritized in green finance, and such an approach is supported by using selective financial tools (green loans and green bonds). The role of green finance is to lead capital towards projects with positive environmental influence. The green finance regulatory environment is changing, with taxonomies, disclosure frameworks, and standard definitions being developed to improve transparency and fight greenwashing. Technological innovation, combined with regulation, especially in digital finance and blockchain, creates new chances and risks for the development of green finance. The Amazon case study is an example of how green finance principles are used in practice by a large company and how initiatives such as sustainability bonds, renewable energy investments, and clean transportation could have a large-scale impact. Nevertheless, it highlights the difficulties and objections companies might encounter in their quest for sustainability. In a world with an urgent need to address climate change and environmental degradation, green finance becomes an important tool in the journey towards a more sustainable future. When financial sources and environmental aims are connected, and innovation in clean technologies is promoted by green finance, it can create an epic win-win from a global perspective. However, tapping this potential involves a joint effort from policymakers, financial institutions, corporations, and investors to create comprehensive frameworks, stimulate transparency, and maintain the credibility of green finance initiatives."

Acknowledgements N/A

Competing Interests The author has no conflicts of interest to declare that are relevant to the content of this chapter.

References

Amanatidis G (2021) Combating climate change. Fact Sheets on the European Union. European Parliament. Retrieved from www.europarl.europa.eu. Website https://www.europarl.europa.eu/factsheets/en/sheet/72/combating-climate-change

Amazon (2021) Bond framework Amazon sustainable Amazon sustainable bond framework 2. Retrieved from https://s2.q4cdn.com/299287126/files/doc_downloads/2021/05/Amazon-Sustainable-Bond-Framework.pdf

Amazon (2022a) Driving climate solutions. Retrieved from Sustainability (US) website https://sustainability.aboutamazon.com/climate-solutions

Amazon (2022b) Sustainability—US. Retrieved from Sustainability—US website https://sustainability.aboutamazon.com/

Andreeva OV, Vovchenko NG, Ivanova OB, Kostoglodova ED (2018) Chapter 2 green finance: trends and financial regulation prospects. Contemp Stud Econ Financial Anal: 9–17. https://doi.org/10.1108/s1569-375920180000100003

Barua S (2019) Financing sustainable development goals: a review of challenges and mitigation strategies. Bus Strat Dev 3(3). https://doi.org/10.1002/bsd2.94

Berrou R, Ciampoli N, Marini V (2019) Defining green finance: existing standards and main challenges. Palgrave Stud Impact Fin: 31–51. https://doi.org/10.1007/978-3-030-22510-0_2

Bhutta AI, Ullah MR, Sultan J, Riaz A, Sheikh MF (2022) Impact of green energy production, green innovation, financial development on environment quality: a role of country governance in Pakistan. Int J Energy Econ Policy 12(1):316–326

Bhutta US, Tariq A, Farrukh M, Raza A, Iqbal MK (2022) Green bonds for sustainable development: review of literature on development and impact of green bonds. Technol Forecast Soc Chang 175(1):121378. https://doi.org/10.1016/j.techfore.2021.121378

BloombergNEF (2020) Sustainable debt sees record issuance At $465Bn in 2019, up 78% from 2018. Retrieved from BloombergNEF website https://about.bnef.com/blog/sustainable-debt-sees-record-issuance-at-465bn-in-2019-up-78-from-2018/

Carè R, Trotta A, Rizzello A (2018) An sustainable finance approach for a more sustainable financial system. Palgrave Stud Sustain Bus Assoc Future Earth: 17–63. https://doi.org/10.1007/978-3-319-66387-6_2

Climate Bonds Initiative (2021a) China green bond market report 2021 climate bonds initiative 1. Retrieved from https://www.climatebonds.net/files/reports/cbi_china_sotm_2021_0.pdf

Climate Bonds Initiative (2021b) Record $269.5bn green issuance for 2020: late surge sees pandemic year pip 2019 total by $3bn. Retrieved March 23, 2024, from Climate Bonds Initiative website https://www.climatebonds.net/2021/01/record-2695bn-green-issuance-2020-late-surge-sees-pandemic-year-pip-2019-total-3bn#:~:text=After%20a%20strong%20first%20quarter

Dervi UD, Khan A, Saba I, Hassan MK, Paltrinieri A (2022) Green and socially responsible finance: past, present and future. Managerial Fin 48(8). https://doi.org/10.1108/mf-11-2021-0561

Díaz-Peña LC, Delgadillo VMC, Iván C-VM (2022) Financial firm's performance: a comparative analysis based on ESG metrics and net zero legislation. J Sustain Fin Invest: 1–21. https://doi.org/10.1080/20430795.2022.2119830

Dorofeev ML (2020) Distinctive features of cost of capital on green bond market. Eco 5(551):62

Dursun-de Neef Ö, Ongena S, Tsonkova G (2023) Green versus sustainable loans: the impact on firms' ESG performance. Retrieved from papers.ssrn.com website https://papers.ssrn.com/sol3/papers.cfm?abstract_id=4115692

Fang Y, Shao Z (2022) Whether green finance can effectively moderate the green technology innovation effect of heterogeneous environmental regulation. Int J Environ Res Public Health 19(6):3646. https://doi.org/10.3390/ijerph19063646

Fu W, Irfan M (2022) Does green financing develop a cleaner environment for environmental sustainability: empirical insights from association of southeast asian nations economies. Front Psychol: 13. https://doi.org/10.3389/fpsyg.2022.904768

Gao Y, Gao X, Zhang X (2017) The 2 C global temperature target and the evolution of the long-term goal of addressing climate change—from the United Nations framework convention on climate change to the Paris agreement. Engineering 3(2):272–278

Gilchrist D, Yu J, Zhong R (2021) The limits of green finance: a survey of literature in the context of green bonds and green loans. Sustainability 13(2):478. https://doi.org/10.3390/su13020478

Gleeson JF (2023) The challenges and opportunities of a net-zero supply chain for Amazon.com

Global Sustainable Investment Review 2020 (2021). https://www.gsi-alliance.org/wp-content/uploads/2021/08/GSIR-20201.pdf

Han Y, Li Y, Wang Q (2023) Digital finance, environmental regulation, and green development efficiency of China. Front Environ Sci 11. https://doi.org/10.3389/fenvs.2023.1131058

Hu Y, Dai X, Zhao L (2022) Digital finance, environmental regulation, and green technology innovation: an empirical study of 278 cities in China. Sustainability 14(14):8652. https://doi.org/10.3390/su14148652

ICMA (2024) Green bond principles. ICMA. Retrieved from www.icmagroup.org website https://www.icmagroup.org/sustainable-finance/the-principles-guidelines-and-handbooks/green-bond-principles-gbp/#:~:text=The%20Green%20Bond%20Principles%20(GBP)
Khan S, Akbar A, Nasim I, Hedvičáková M, Bashir F (2022) Green finance development and environmental sustainability: a panel data analysis. Front Environ Sci 10. https://doi.org/10.3389/fenvs.2022.1039705
Lalon RM (2015) Green banking: going green. Int J Econ Fin Manage Sci 3(1):34. https://doi.org/10.11648/j.ijefm.20150301.15
Liu C, Wu SS (2023) Green finance, sustainability disclosure and economic implications. Fulbright Rev Econ Policy 3(1). https://doi.org/10.1108/frep-03-2022-0021
Liu H, Zhu Q, Khoso WM, Khoso AK (2023) Spatial pattern and the development of green finance trends in China. Renewable Energy 211:370–378
Neumann M (2023) The political economy of green bonds in emerging markets. Retrieved March 23, 2024, from springerprofessional.de website https://www.springerprofessional.de/en/the-political-economy-of-green-bonds-in-emerging-markets/25369792
Ozili PK (2022) Green finance research around the world: a review of literature. Retrieved from papers.ssrn.com website https://papers.ssrn.com/sol3/papers.cfm?abstract_id=4066900
Pagitsas C (2022) Kara Hurst: Amazon. In: Chief sustainability officers at work: how CSOs build successful sustainability and ESG strategies. Apress, Berkeley, CA, pp 103–117
Raimi D, Campbell E, Newell R, Prest B, Villanueva S, Wingenroth J (2022) Global energy outlook 2022: turning points and tension in the energy transition. Resources Future, Washington, DC, USA
Rosenblatt L (2023) Does Amazon's sputtering push to be a better, greener employer matter? Retrieved from The Seattle Times website https://www.seattletimes.com/business/does-amazons-push-to-be-a-better-greener-employer-matter-to-markets/#:~:text=Internally%2C%20Amazon%20has%20committed%20to
Schumacher K (2023) Environmental, social, and governance (ESG) factors and green productivity: the impacts of greenwashing and competence greenwashing on sustainable finance and ESG investing. SSRN Electron J. https://doi.org/10.2139/ssrn.4303609
Taghizadeh-Hesary F, Yoshino N (2019) The way to induce private participation in green finance and investment. Financ Res Lett 31:98–103. https://doi.org/10.1016/j.frl.2019.04.016
TCFD (2021) Task force on climate-related financial disclosures. Retrieved from TCFD website: https://www.fsb.org/uploads/P141021-1.pdf
TCFD (2023) Task force on climate-related financial disclosures. Retrieved from TCFD website https://www.fsb-tcfd.org/
Wang T, Liu X, Wang H (2022) Green bonds, financing constraints, and green innovation. J Clean Prod 381:135134. https://doi.org/10.1016/j.jclepro.2022.135134
Zhang D (2023) Does green finance really inhibit extreme hypocritical ESG risk? A greenwashing perspective exploration. Energy Econ 121:106688–106688. https://doi.org/10.1016/j.eneco.2023.106688
Zhao L, Chau KY, Tran TK, Sadiq M, Xuyen NTM, Phan TTH (2022) Enhancing green economic recovery through green bonds financing and energy efficiency investments. Econ Anal Policy 76:488–501

Understanding Green Bonds: A Comparative Analysis of India and China as Pioneers in Climate Financing

Saurabh Sood and Samreedhi Gupta

Abstract The advent of industrialisation has expedited the process of environmental degradation, resulting in an urgent need to devise innovative solutions to facilitate sustainable development. With a tripartite focus on ecological integrity, societal equity, and economic feasibility, sustainable development has become a harbinger of global progress, indicating how companies significantly influence the environmental, social, and economic ethos. Green Bonds is a novel form of finance designed to advance sustainable investing in environmentally friendly projects. Originating in the European Union in 2008, they still exist at a nascent stage in the Indian subcontinent. While the Companies Act 2013 has attempted to harmonise corporate operations and sustainable development in India, much is yet to be accomplished. India's sovereign green bonds are majorly unregulated, making its commitment to sustainable development a farce. The present research attempts to delve deeper into the symbiotic and multifaceted relationship between green bonds as a climate financing mechanism and the Companies Act 2013 to underscore how the legislative framework acts as a catalyst for advancing conscientious business processes. It employs a comparative method for investigating the evolving interplay of green bonds in the Indian context. It concludes with a suggestive guide for governing green bonds in India.

Keywords Climate financing · Green bonds · Sustainable development · India · China

S. Sood (✉) · S. Gupta
Jindal Global Law School, O.P. Jindal Global University, Sonipat, Haryana 131001, India
e-mail: saurabhsood78@gmail.com

S. Gupta
e-mail: guptasamreedhi@gmail.com

H. Baber and M. Fanea-Ivanovici (eds.), *Sustainable Financing—A Contemporary Guide for Green Finance, Crowdfunding and Digital Currencies*, World Sustainability Series, https://doi.org/10.1007/978-3-031-80969-9_3

1 Introduction

Internationally, no universally accepted definition of 'green bond' exists, and industry associations, including the International Capital Markets Association (ICMA) (Weber and Saravade 2019a, b), and stock exchanges, have put forth requisites for a bond to qualify as a green bond. Chief among the requirements is the need for an agreement to utilise the cash proceeds of the green bonds to finance new or existing projects that benefit the environment and climate (Carvalho 2018). Overall, green bonds are financial debt instruments wherein the issuer promises to repay an amount to the holder with interest on a specific future date, and such proceeds shall be utilised for environmentally friendly financing projects (Tuhkanen 2020). In a way, green bonds provide investors with income payments that are either fixed or regular (Karthik 2022). For over a decade now, climate change and associated challenges have been addressed with the help of green bonds. Such challenges include threats to communities and economies and present palpable hazards to agriculture, water, resources, etc (Flammer 2021). A reasonable amount of investments must be channelled concertedly to redress the issues by aligning projects, investors, capital, and sustainable development. Green bond affords a golden thread that draws these factors together (Panigrahi and Banerjee 2021). With a green financing market value of around seven billion dollars, India is ranked at number eight in financing "green" projects (Patidar 2023).

The study locates itself within the context of sustainable development in India to explore the symbiotic relationship between green bonds as a tool for climate financing and Companies Act 2013, while using China in a comparative study. The experience of China is important because out of the total green bond proceeds of its exponential rise in the past decade in the global green bond market. Unlike the centralised system of governance to regulate the green bonds in other countries, the Chinese government has a strong control over green bond issuance in China. To this end, a top-down approach is used by the sector specific regulators and brings more clarity in the issuance of the green bonds. A comparative analysis of the diverse approaches followed in India and China, helps in understanding Green Bond for sustainable outcome. In doing so, the authors critique the present green bond regulations in India and recommends adopting a standard green bind framework based on China's top-down and sector-specific approach.

2 Background

In the paper *Turning Billions into (Green) Trillions*, the author emphasises that the green bond market is likely to grow by leaps and bounds and will prove to be a decisive instrument for climate change mitigation. For augmenting green bonds in the market, needs the cooperation of different stakeholders such as central banks, Regulators in finance, countries, and the business community in particular. On one hand countries such as China, France, India have been at the forefront of the green bond space, not much work has been done by other countries. Similarly United States and its corporations have made significant developments in this space and has emerged has a pivotal actor in the global bond market and has the potential to lead the marketplace. Hence, the policymakers across the world need to harness the prospect of green bonds and transition into a green economy (Faske 2018).

In the paper *Transforming finance for sustainability in energy: past, present and future of green bonds in China*, the author Xu Qian discusses the current regulatory framework of the Chinese green bond market. While highlighting the significant efforts made by the Chinese government to attract international investors and make China a leading player in the global green bond market, Qian lauds the sector-based regulatory mechanism adopted by it and presents a critique of the system to the multiply the progress of green bonds in the Chinese energy market (Qian 2022).

Leah Temper and Joan Martinez-Alier, in their article *Is India Too Poor to Be Green?* point out that it may is possible to convert the environmental services to monetary values in the form of carbon credits, yet there are certain aspects of it, which still can't be translated in monetary terms. India has transitioned from an agrarian society to an industrial society has been gradual and it has not replicated the west in this process. The authors emphasise that clean air and rainfall cannot be treated as commodities and be bought and sold in the market. Destruction of local cultures and biodiversity attempting to quantify them in monetary terms is incommensurable. Initiatives affecting environment have social costs which are irreversible and similarly once a particular region is exploited and the minerals are all gone, there is no going back. The paper is important because it helps in carefully understanding the costs of tampering with environment and further treading carefully while dealing with issuance of green bonds (Temper and Martinez-Alier 2007).

The paper *A Climate Handshake- The India-US Green Strategic Partnership*, by Pramit Pal Chaudhuri, an attempt has been made to look at climate change from a India-US perspective. According to author, the world is facing the immediate threat to reduce the carbon emissions because its increase is leading to irreversible and drastic changes. The paper underscores, the need for requisite funding for India to achieve its climate goals. While the government of India has identified the costs, yet the paper argues that much of the expenses will be raised from the private capital markets and mostly in the form of debt and a small portion in the form of equity. Further, it is unrealistic to expect the private players to infuse money in India's solar power projects or speculative projects such as the electrical vehicles. The green bond market in India is underdeveloped and the predominant market for Indian firms to raise green bonds is the London Stock Exchange and must incur not only hefty interest rates but also steep underwriting fees. In 2018 just US$7 billion worth Indian green bonds were raised which is a fraction of what of what countries like US and China raised. Therefore, a lot of path needs to be covered to make green bonds more viable commercial product (Chaudhuri 2020).

Olaf Weber and Vasundhara Saravade, in their paper *The Growth of Green Bonds* highlight how green bonds are changing the way societies are financed especially from the perspective of Socially Relevant Investment. Although the growth of green bonds is a positive, yet it is still not clear whether additional investments are attracted by green bonds, or they are a rehash of the conventional bond issuances which would anyways by issued. The paper reiterates the seminal idea that the conventional bond market needs to be green and climate change and other environmental issues need to be addressed. Using the empirical data, the author indicates that there has been an increase in the green bonds absolute value, and this has led to increase in issuers and investors in the green bond market each year. This has given rise to a greater need for accountability and transparency so that the issues of greenwashing are mitigated. Greenwashing in terms of the green bond market indicates a situation wherein the proceeds from the green bonds get allocated to assets bearing little or no environmental value. Therefore, certified green bonds are useful for conventional investors and SRI investors (Weber and Saravade 2019a, b).

3 Legal Framework for Green Bonds: Navigating India's Regulatory Landscape

In India, the Securities and Exchange Board of India ("SEBI") issued a comprehensive Circular on Disclosure Requirements for Issuance and Listing of Green Debt Securities (Disclosure Requirements), 2017 (SEBI 2017), which necessitates an issuer of a green bond or a debt security instrument to furnish certain disclosures (Manaktala 2020), such as:

1. The environmental objective of issuance of such a bond.
2. Information on the procedure to be used for keeping track of the utilisation of the proceeds.
3. The issuer sets up particulars of assets or projects using the proceeds from the green debt securities.

Further, the SEBI (Issue and Listing of Non-Convertible Securities) regulations, 2021 explain green debt security as a debt security that is issued to raise funds that are to be spent on projects or assets that come under the following:

1. Renewable energy (including wind, solar, bioenergy, and any other energy sources using clean technology);
2. Clean transportation;
3. Sustainable water management systems;
4. Energy efficient and green buildings;
5. Climate change adaptation
6. Sustainable waste management;
7. Biodiversity conservation; and
8. Any other category specified by the SEBI from time to time.

The Constitution of India envisages sacrosanct principles for protecting the environment, forests, wildlife, etc. In pursuance of these cherished goals, the Government of India in the year 2008 launched the National Action Plan on Climate

Change, intending to fulfil eight national missions entailing the reduction of emissions, ensuring energy efficiency, forest cover, and creating sustainable habitats. The policy for climate protection is part of the various other policy goals concerning energy access and water conservation (Haran 2023). Considering the importance of sustainability, even Regulation 34 of the SEBI (Listing Obligations and Disclosure Requirements) Regulations 2015 obligates the top 1000 entities based on the market capitalisation to furnish a thorough annual report containing Business Responsibility and Sustainability Report (BRSR) mentioning the ESG programs undertaken by the company. Following the Green Bond Principles, the measures taken by SEBI in the form of a legal framework are also available on the SEBI website.

Similarly, on July 12, 2023, SEBI introduced the much-awaited BRCR Core, which forms part of the larger BRSR framework. The underlying purpose is objectively establishing measurable criteria to report the positive principles enumerated under the BRSR. This has ensured a shift towards more outcome-oriented and measurable Environmental, Social, and Governance benchmarks. Apart from the regulatory measures, the Companies Act 2013, with its extant rules, specifies green CSR activities, their reporting, and assessments to advance best corporate governance practices.

The market for green bonds in India comes under the corporate bond market. To ensure that the green bonds receive their due recognition, it is crucial for the bond market to also go through reforms (Lee 2015). As previously discussed, in the Circular on Disclosure Requirements for Issuance and Listing of Green Debt Securities issued in 2017, any debt security is to be considered a green or green debt security if the funds from such an issuance can be used for either of the eight categories enumerated above. Therefore, under the said Circular, particular disclosures are required to be made (Anand 2022):

1. Details of the project or assets.
2. Environmental objective of the issue.
3. Details about the monitoring of disbursement and usage of funds.

Such disclosures are regular and periodic and must be submitted with annual reports and related financial statements. To certify the entire process and evaluation of the project, the issuer may appoint an independent third-party certifier and reviewer. In a nutshell, the issuance and listing of green bonds or green debt securities are subject to the following and are collectively referred to as the Green Debt Securities Regulations:

1. Companies Act, 2013
2. SEBI (Issue and Listing of Debt Securities) Regulations, 2008
3. SEBI (Listing Obligations and Disclosure Requirements), 2015.
4. Circular on Disclosure Requirements for Issuance and Listing of Green Debt Securities issued in 2017
5. SEBI (Issue and Listing of Non-Convertible Securities) Regulations, 2021.

Reserve Bank of India secured a membership in the Network for Greening the Financial System on April 23, 2021, to learn and contribute to the global efforts on green finance (Rao 2021). India issued its first sovereign green bonds worth Rs 80 billion, which equates to $968 million in January 2023 (World Bank Case Study 2023).

4 Regulatory Complexity in the Green Bond Sphere: Exploring the Possible Challenges in India

As per the Fitch Ratings report, the market for green bonds in India, compared to the global green bond market, stood at a mere 1.7%, which amounted to $8.6 billion. The three reasons that have prevented the growth of green bonds are: firstly, the scope of the issuance of green bonds is limited to the renewable energy sector, specifically solar projects (Tiwari and Singh 2023). Secondly, the issuers and the investors are not incentivised enough, and thirdly, the regulatory framework to regulate green bonds is not exhaustive. Reserve Bank of India, through its working paper 'Green Finance in India: Progress and Challenges', revealed that of all the bonds issued in India, green bonds make up only 0.7% (Jain and Gangopadhyay 2022).

1. *Promoting Uniformity in Green Bond Standards and Certification*

In addition to the above guidelines at national level, the green bonds and its credibility is strengthened by certified standards (Bongaerts and Schoenmaker 2020). Climate Bonds Initiative (CBI) which is a London based organisation which is regarded as a pivotal agency for certifying the green bonds by issuing Climate Bond Standards. Whenever a green bond is issued, in order to procure a Climate Bond Certification Stamp, it is mandatory that an approved third party verifier is appointed by the prospective issuer of the green bond. This third party verifier, is then required to provide a verification statement certifying that the bond complied with the Climate Bonds Initiative 2019 credentials. In India, Hero Wind Energy green bond worth $44 million was the first certified green bond to receive a third-party verification by KPMG and CBS in January 2016 (NRDC Draft Report 2016). Adopting international standards has been subjected to criticism since it may not apply to the domestic context and therefore the need to have a more domestic approach is required. One suggestion here would be to strengthen the domestic credit rating agencies namely, CARE Ratings Limited, Credit Rating Information Services of India Limited (CRISIL) and ICRA Limited. The purpose can be achieved by tailor making the certification standards to align them to the Indian sensibilities and at the same time affirming that the international credibility of the domestic green bonds is not misplaced (Oche 2020).

2. *Absence of authoritative precedent*

Due to its unwavering stand and commitment towards financial practices that are environmentally sustainable, the Supreme Court of India has exhibited a positive standpoint for green financing. Irrespective of acknowledging the importance of green financing, it is worth noting that the Supreme Court of India has not delivered any authoritative judgment on green bonds. There is a conspicuous absence of a decisive legal precedent that clarifies green bonds. This absence of judicial pronouncement on green bonds may be understood to be a gap in the comprehensive legal framework and leaves room for much scope for developing jurisprudence around green bonds (Mori and Shimizu 2019).

3. *Issuance of green bonds is disincentivised to a reasonable extent*

At the core of the debate surrounding green bonds is the lack of 'greenium,' which is essentially the difference in the financial benefit from a green bond compared to conventional bonds. Considering the long-term implications, it is highly improbable that any investor will accept lower returns through green bonds. A possible answer to the lower interest rates of green bonds could be offering tax incentives to the government. Similarly, under the Income Tax Act 1961, certain tax exemptions may be applicable upon income arising from investments in green bonds.

Currently, green bond investors are not allowed to invest in supporting ongoing projects, and as a consequence, developers seek loans to fund their projects, not through green bonds. On the other hand, the United States has a formalised incentive approach within the Clean Renewable Energy Bonds (CREBs) and Qualified Energy Conservation Bonds (QECBs). Green bonds have been classified into three types, i.e. tax-exempt bonds, tax credit bonds, and direct subsidy bonds, which offer a prospective investor, issuer, or both with a diverse monetary incentive.

4. *The absence of credit rating agencies in the domain of green bonds market*

There is an absence of credible ratings for green bonds or projects, which affects the creditworthiness of these projects to substantial projects (Holmes 2010).

5. *Systemic issues*

The commercial interests of the issuers have been at the receiving end due to the systemic issues in the green bond market in India. For instance, the issuers of green bonds are not given any pricing advantage because developer companies involved in renewable energy are responsible for 80% of the total issuance of green bonds. A glaring issue with companies engaged in renewable energy is that their clients are the discoms or the distribution companies run by different states.

These clients are financially weak, resulting in issuers being rated poorly and increasing interest rates levied by the investors (Yu 2018). This proves less profitable for the developers who issue such green bonds. Green bonds are held by the Reserve Bank of India, issuers, or domestic banks and are expressed in Indian rupees. However, around 90% of the issuers opt for dollars while issuing green bonds, which causes reduced interest for the issuers. Lastly, since the green bonds are issued in Indian rupees, they directly thwart foreign investments in the Indian green bond space (Sean 2019).

6. *Greenwashing*

The proceeds of the green bonds are to be used for specific purposes. The reality is, however, different because the proceeds are used for projects with zero to almost negligible effect on the environment (Lin 2023). This practice is called greenwashing and makes the whole issuance process of green bonds a farce.

7. *Inadequacy of legal framework*

The legal framework for green bonds has been put forth comprehensively recently, with SEBI releasing the Revised Disclosure Requirements for the issuance and Listing of Green Debt Securities in February 2023. This move has infused transparency and some clarity in the process by making the publication of a statement mandatory on due diligence policies on adverse investment results. The disclosure entails information furnished through annual periodic reports by the issuers of these bonds, which then get examined by the third-party certifiers. Though constructive, these rules still lack clarity on the standardisation and certification of corporate green bonds (CGBs).

The lack of practical rules for standardising and certifying green bonds makes them less credible for global and domestic investors (Ranaut 2021). A possible solution to this can be taken from the European Union, where an additional framework to support the issuance of green bonds referred to as the European Green Bond Standard, which ascribes the designation of European Green Bond on such bonds that fulfil specific requirements such as usage of the proceeds of the European Green Bond for the economic activities which conform with the EU taxonomy requirements on sustainable finance. Another important legal document is the Corporate Sustainability Reporting Directive (CSRD), which imposes a duty upon the parties issuing the European Green Bonds to initiate the audited transition plans to ensure reduced carbon emissions. However, this requirement of preparing a transition plan is only applicable to large businesses, not small and medium companies (Wang 2018).

8. *Added costs*

As discussed earlier, compared to other conventional bonds in India, the issuance of green bonds has proved to be pricey. Various factors can be attributed to it, and chief among them is the absence of a platform for the reporting, verifying, and tracking of green bonds or climate finance generally. Similarly, inadequate policy framework at the national and state level exacerbates the issue.

9. *Brief tenure coupled with indefinite returns*

Once the green bonds are issued in India, they have a shorter tenure of approximately 10 years in contrast to a 13-year tenure of a general loan. Similarly, green projects often take up a lot of years before finally yielding some results, financial or otherwise.

5 China's Legal Framework for Sustainable Finance: A Focus on Green Bonds

Since issuing its first green bond in 2015, China has established itself as an instrumental player in the global green bond market (Maino 2022). Being the second largest in the world (Christian and Zhang 2019), the green bonds structure in China is deemed one of the most rigorous in the world (Rivera 2022). It comprises three regulatory protocols, namely:

1. The People's Bank of China Announcement, 2015 ("PBoC") provides the fundamental regulatory measures of green bonds in China. (Tuhkanen 2020). It enlists the projects that may be funded with green bonds under the Green Bonds Endorsed Projects Catalogue ("EPC"), which categorises these projects as (a) Clean Energy, (b) Clean Transport, (c) Resource Conservation and Recycling, (d) Pollution Prevention and Control; (e) Energy Conservation; and (f) Ecological Protection and Climate Change Adaptation (PBOC 2015).
2. National Development and Reform Commission's Guidelines on Green Bond Issuance in 2015 ("NDRC") states requirements and streamlined approval procedures along with the provisions on the issuance of corporate green bonds (Chen and Zhao 2021)
3. China Securities Regulatory Commission's Guiding Opinions for Supporting Green Bonds, 2017 ("CSRC") lays down the broader framework for regulating green bond trade through stock exchanges. It released standards and qualifications for ratings and verification.

In addition to the aforementioned essential regulations related to green bonds in China, the National Association of Financial Market Institutional Investors Opinions ("NAFMII") introduced the Guidelines on Green-Debt Financing instruments for non-financial enterprises (Galagher and Qi 2016). Banking institutions and state-owned enterprises primarily dominate the market, and green bonds are predominantly issued among these entities, where commercial banks purchase and sell bonds in favour of investors for long and short terms (Chen and Long 2023). Additionally, the high credit rating of green bonds in China makes them a risk-free asset, having enough buyers in the market at all times (Cheng and Drahos 2022).

Although there is no universally accepted criteria for funding projects through green bonds, international standards suggest that green bonds must be used for green projects exclusively. However, a close examination of the Chinese economy indicates that it allows the allocation of 10% of the total green bond proceeds to general corporate purposes such as clean coal projects instead of green projects (Wang et al. 2020). The EPC, which identifies categories for green bond investors, lays down standards related to the eligibility of sub-sectors that suit the needs of financing the Chinese economy and balancing the interests of a green economy simultaneously.

Another notable deviation from the international norm is the type of governance employed in China concerning green bonds. Unlike the decentralised system of governance for the regulation of green bonds in other states, the Chinese government exercises strong influence and regulatory oversight of green bond issuance (Park 2018). The concerned sector-based regulators especially promulgate the top-down approach followed by the government (Lin and Hong 2022). Therefore, the national legislation framed by the National People's Congress allows the diversification of issuing multiple green bonds by providing sector-based guidelines that must be read in accordance with the national provisions, thereby facilitating a smoother process for issuing green bonds such that they are tailored according to the requirements of that particular sector (Zhang 2020). The sector-based approach has also clarified the process, increasing the issuance of green bonds (Zhang 2021).

6 Harmonising Legal Landscapes: Recommendations for India Based on China's Green Bond Framework

China's green bond structure is primarily operated by four government agencies, including the PBoC, NDRC, CSRC, and NAFMII. These organisations mandate the disclosure of essential information related to the bonds- eligibility of green bonds, targeted environmental benefits and progress of projects towards realising the goal sought to be achieved (Verma and Agarwal 2020). These relatively strict standards dominated by state-owned enterprises and regular monitoring of developments in the green bond market have paved the way for a steady growth of the green bond market in the nation. Thus, developing countries and emerging green bond markets like India must emulate China's strategies to envisage a better ecosystem for green bonds.

1. *Introduce uniform standards*

The disclosure of important information is accompanied by third-party certification following the NAFMII Guidelines in China. Although it is yet to become a legally binding requirement, obtaining a recognised certification during the different stages of issuance makes the entire green bond structure transparent. Moreover, it restores confidence in the system (Prakash and Sethi 2021). The Standard and Certification Scheme allows stakeholders to determine a commercial project's "green" element and prioritise climate-friendly investments (Ehlers and packer 2017). Although India recognises outside certification (Lindenberg 2014), it lacks credibility, which has led to a decline in investments made toward green bonds in the sub-continent (Bansal 2020).

2. *Strengthen Incentives for the issuance of green bonds*

Unlike India, the green bond market has less credit and liquidity risk than other market-oriented nations in China (Qian 2022). Since the system mainly consists of banks and public-sector undertakings, the overall credit rating is high, and chances of default are low, making green bonds a viable and risk-free investment for investors, thereby incentivising more people to invest in the Chinese green bonds market as an excellent security in their portfolio (Su and Zhang 2022).

3. *Transparency*

Apart from several disclosures during the issuance of green bonds, the NAFMII mandates to disclose the level of "green" of debt financing instruments and implementing regular follow-up assessments to ensure compliance with eco-friendly projects, preventing instances of greenwashing in the economy and securing the faith of investors and their money towards environment-friendly practices (Cao et al. 2021).

7 Conclusion

India has been working towards financing green projects effectively and efficiently in the last few years. In pursuance of these cherished goals, environmentally sustainable practices have been introduced. These concerted steps have aided in accomplishing carbon neutrality to a reasonable extent. Green bonds are pivotal in climate finance but face many structural issues. In other words, though the government and the Regulators in India have put into effect many measures to support green financing by streamlining the legal framework of green bonds, much work still needs to be done. Therefore, as the market continues to grow, there is a need to analyse the regulatory experiences of other jurisdictions to gain practical insights of these nations.

A probable solution is the inclusion of a guaranteed mechanism for the creditors giving assurance whereby, in any unforeseen event, a third entity, if a borrower cannot make any payment due to not receiving the dues from its clients, i.e. discoms. At times, blanket guarantees are expensive; hence, a partial guarantee could be a possible solution, provided they don't charge exorbitant fees from the developers. Secondly, offering diverse and incentivised green bonds in the Indian market such as the standardised framework offered in the European Green Bond Standard, can prove to be a financially viable option as this would help green bonds compete with the more conventional investment instruments in India. For instance, the inclusion of tax incentives on the revenue generated from green bonds under the Income Tax Act should also be introduced. The low costs of issuance of green bonds will augment investor confidence and help the government achieve the Nationally Determined Contributions goals, which are at the heart of the Paris Agreement (Nagpal and Jain 2018). Thirdly, since India lacks a standard corporate green bond framework, help may be taken by following China's top-down approach and sector-based regulations. The framework is not only equipped with flexibility to suit sector-specific circumstances, but also streamlines the process of issuance of green bonds through the enactment of a national legislation. Finally, harmonisation of national standards for green bond issuance is a necessity to develop the green bond market in India as it aids in securing the confidence of investors and cementing its place as a sustainable and viable investment (Weber and Saravade 2019a, b).

8 Competing Interests

The authors have no conflicts of interest to declare that are relevant to the content of this chapter.

References

Anand A (2022) The effectiveness of green bonds in India: a socio-legal perspective. DME J Law 3(01):77–85. https://doi.org/10.53361/dmejl.v3i01.11

Bansal S (2020) Green bonds-trend and challenges in India. J Bus Manage Inf Syst (7)1:22–30. https://doi.org/10.48001/jbmis.2020.0701003

Bongaerts D, Schoenmaker D (2020) Green certificates: a better version of green bonds. Bruegel Policy Contribution No. 2020/20

Cao X, Jin C, Ma W (2021) Motivation of Chinese commercial banks to issue green bonds: financing costs or regulatory arbitrage? China Econ Rev 66:101582. https://doi.org/10.1016/j.chieco.2020.101582

Carvalho G (2018) Climate finance for Canadian cities: is debt financing a viable alternative. Institute of Municipal Finance and Governance

Chaudhuri PP (2020) A 'climate handshake': the India-US green strategic partnership. Indian Foreign Affairs J 15(4):328–334

Chen X, Long W (2023) To enhance the credibility of the green bond market through regulating GBERs: the case of China. Laws 12(6):91. https://doi.org/10.3390/laws12060091

Chen Y, Zhao ZJ (2021) The rise of green bonds for sustainable finance: global standards and issues with the expanding Chinese market. Curr Opin Environ Sustain 52:54–57

Cheng W, Drahos P (2022) China's regulatory pathway to green finance. Tsinghua China Law Rev 14(2):195–219

Christian E, Zhang LY (2019) Diffusion and innovation for transition: transnational governance in China's green bond market development. J Environ Planning Policy Manage 21(4):391–406. https://doi.org/10.1080/1523908X.2019.1623655

Ehlers T, Packer F (2017) Green bond finance and certification. Bank Int Law Settlement Q Rev: 89–104

Faske B (2018) Turning billions into (Green) trillions: tracking the growth and development of the green bond market in China, France, India, and the United States. Tulane Environ Law J 31(2):293–325

Flammer C (2021) Corporate green bonds. J Fin Econ

Galagher KS, Qi Q (2016) Policies governing China's overseas development finance implications for climate change. Centre for International Environmental and Resource Policy, The Fletcher School

Haran C (2023) Political economy of green finance: analysing its effect on the development of legal framework. Indian J Integr Res Law 3(3)

Holmes I (2010) Green bonds: The missing piece of the Jigsaw. E3G

Jain K, Gangopadhyay M, Mukhopadhyay K (2022) Prospects and challenges of green bonds in renewable energy sector: case of selected Asian economies. J Sustain Fin Invest: 1–24. https://doi.org/10.1080/20430795.2022.2034596

Karthik KP (2022) Securing green bonds with carbon credits. Jus Corpus Law J: 303–400

Lee A (2015) DEAL: India Exim green bond precedent. Int Fin Law Rev 34(19):11

Lin JD (2023) Explaining the quality of green bonds in China. J Clean Prod: 136893. https://doi.org/10.1016/j.jclepro.2023.136893

Lin L, Hong Y (2022) Developing a green bonds market: lessons from China. Euro Bus Org Law Rev 23(1):143–185

Lindenberg N (2014) Definition of green finance. German Development Institute. https://www.die-gdi.de/uploads/media/Lindenberg_Definition_green_finance.pdf

Maino AG (2022) Green bonds: a global perspective. In: Financing the energy transition: the role, opportunities and challenges of green bonds. Oxford Institute for Energy Studies, 3–11

Manaktala S (2020) Green bonds in sustainable finance: exploring the case of India. NITI Aayog. https://doi.org/10.2139/ssrn.3644116

Mori N, Shimizu N (2019) Current situation and challenges of green bonds in japan from the view point of promoting repeat issuance.: analysis of results of a questionnaire survey to issuers. Institute for Global Environmental Strategies. https://doi.org/10.57405/iges-705657405/iges-7056

Nagpal A, Jain M (2018) 'Green is the new normal in the sustainable era': special reference to green bonds market in India. TERI Inf Digest Energy Environ 17(3):285–298

NRDC Draft Report (2016) Greening India's financial market: how green bonds can drive clean energy deployment. https://www.nrdc.org/sites/default/files/india-financial-market-green-bonds-report.pdf

Oche A (2020) Comparative analysis of green bond regimes in Nigeria and China. J Sustain Dev Law Policy 11(1):160–184

Panigrahi B, Banerjee H (2021) Development of green FDI in India. Ind J Law Legal Res 2(1)

Park SK (2018) Investors as regulators: green bonds and the governance challenges of the sustainable finance revolution. Stanford J Int Law 54(1)

Patidar M (2023) Green bonds and sustainable finance: analyzing market trends and impact on environmental initiatives in India. Universal Res Rep 10(4):186–192

Prakash N, Sethi M (2021) Green bonds driving sustainable transition in Asian economies: the case of India. J Asian Fin Econ Bus 8(1):723–732

People's Bank of China (2015) Preparation instructions on green bond endorsed project catalogue, 2015 edn. https://policy.asiapacificenergy.org/sites/default/files/PBOC-Announcement-No-39-2015.pdf

Qian X (2022) Transforming finance for sustainability in energy: past, present and future of green bonds in China. J World Energy Law Bus 15(5):319–332. https://doi.org/10.1093/jwelb/jwac017

Ranaut V (2021) Green financing in India: addressing the challenges ahead. Int J Law Manage Human 4:1779–1788

Rao DT (2021) A study on green bonds in India-need of the hour. Int J Res Anal Rev 8(4):62–69

Rivera B (2022) Green bonds: reforming ESG regulation in the United States to meet the requisite funding demand for a decarbonized economy. Hastings Environ Law J 28

Sean Kidney, C. E. O., & Initiative, C. B. (2019) Moving from growth to development: financing green investment in India. Fin Green Trans 98

Securities and Exchange Board of India (2017) Revised disclosure requirements for issuance and listing of green debt securities, CIR/IMD/DF/51/2017

Su T, Zhang ZJ, Lin B (2022) Green bonds and conventional financial markets in China: a tale of three transmission modes. Energy Econ 113:106200

Temper L, Martinez-Alier J (2007) Is India too poor to be green? Econ Pol Wkly 42(17):1489–1492

The World Bank Case Study (2023) India sovereign green bond: financing climate action and resilient growth. https://thedocs.worldbank.org/en/doc/c68d5c90796897b1628c25fea3590a5b-0340012023/original/Case-Study-India-Green-Bond-TA.pdf

Tiwari D, Singh A (2023) Exploring the role of municipalities in promoting sustainable development with special reference to green bonds in India. Int J Environ Sci 9(2)

Tuhkanen H (2020) Green bonds: a mechanism for bridging the adaptation gap? Stockholm Environment Institute, 8–15

Verma A, Agarwal R (2020) A study of green bond market in India: a critical review. IOP Conf Ser Mater Sci Eng 804(1)

Wang EK (2018) Financing green: reforming green bond regulation in the United States. Brooklyn J Corp Fin Commer Law 12(2):467–491

Wang J, Chen X, Li X, Yu J, Zhong R (2020) The market reaction to green bond issuance: evidence from China. Pac Basin Financ J 60:101294

Weber O, Saravade V (2019a) Green bonds: current development and their future. Centre for International Governance Innovation, Paper No. 210

Weber O, Saravade V (2019b) The growth of green bonds. in green bonds: current development and their future. Centre for International Governance Innovation, 7–11. http://www.jstor.org/stable/resrep24966.12

Yu Y (2018) Catalysing green bonds in ASEAN+3 countries to mobilise finance for energy market transformation. Institute for Global Environmental Strategies

Zhang H (2020) Regulating green bond in china: definition divergence and implications for policy making. J Sustain Fin Invest 10(2):141–156. https://doi.org/10.1080/20430795.2019.1706310

Zhang LY (2021) Conducting and financing low-carbon transitions in China: a governmentality perspective. Edward Elgar Publishing

Green Finance for Sustainable Financing

Rosa Adamo, Domenica Federico, and Antonella Notte

Abstract Sustainable finance takes into account environmental, social and corporate governance factors in the investment decision-making process, directing capital towards sustainable activities and projects. A new model of financial sustainability is developed with the aim of directing capital towards projects aimed at respecting the environment and with an ecological vocation. Green finance emerges within this model. It allows green investments to be favored over traditional ones which, instead, often fuel unsustainable growth models. The main products of green finance can be divided among green credit market instruments and green capital market instruments, and these products have a leading role in the economic system. Green finance increasingly stimulates the transfer of financial resources to economic entities engaged in environmental and eco-sustainable projects and activities. Green finance encourages firms to adopt more environmentally friendly production processes and technologies through green investments. This chapter starts from an overview of green finance as a new model of sustainable finance and then identifies the main green finance instruments and defines their role in encouraging firms to engage in green innovation. Examining the implications and challenges of green finance as a powerful tool for redirecting financial flows towards sustainable investments and green business innovations concludes the chapter.

Author's Contributions The paper is the fruit of the joint work of all the authors. However, Rosa Adamo mainly contributed to Sects. 1 and 6, Domenica Federico to Sects. 3 and 5, and Antonella Notte to Sects. 2 and 4.

R. Adamo
Department of Business and Law, University of Calabria, Arcavacata of Rende, CS, Italy
e-mail: rosa.adamo@unical.it

D. Federico (✉) · A. Notte
Department of Social and Human Sciences, eCampus University, Novedrate, CO, Italy
e-mail: domenica.federico@uniecampus.it

A. Notte
e-mail: antonella.notte@uniecampus.it

H. Baber and M. Fanea-Ivanovici (eds.), *Sustainable Financing—A Contemporary Guide for Green Finance, Crowdfunding and Digital Currencies*, World Sustainability Series, https://doi.org/10.1007/978-3-031-80969-9_4

Keywords Green finance · Sustainable finance · Financial instruments · Business innovations · Decision-making process

1 Introduction

Sustainable finance integrates finance with sustainable development goals, investing in initiatives that meet Environmental, Social and Governance (ESG) criteria. From this perspective, a new model of financial sustainability is developed with the aim of increasingly directing investment decisions towards greater awareness of the environment, people and society. Within this model, attention to environmental issues emerges to meet the long-term needs of an inclusive and sustainable economy. In particular, green finance is defined and refers to a set of financial interventions whose proceeds are used for low-carbon, sustainable and inclusive projects with the aim of promoting environmentally sustainable paths and initiatives. Green finance allows you to give priority to green investments over traditional investments which, instead, often fuel unsustainable growth models.

In the field of green finance, financial guidelines and strategies are defined that guarantee transparency and encourage a long-term vision of investments in compliance with the sustainable development criteria identified by the United Nations Sustainable Development Goals. In particular, at a European level, the European Commission continued the path undertaken of environmental sustainability until the entry into force of the Taxonomy Regulation 2020/852 which allowed the horizons of various companies to be oriented towards greener investments.

Green finance encompasses a wide range of financial products and services, which can be divided into green credit market instruments and green capital market instruments. These instruments play a crucial role in financing projects related to renewable energy, energy efficiency and clean transport, which are essential components of sustainable development. The development of a green finance market has a driving role in the economic system. In fact, on the one hand, it encourages the transfer of financial resources towards economic entities that are engaged in environmental and eco-sustainable projects and activities, on the other hand, it increasingly stimulates firms to adopt more environmentally friendly production processes and technologies through green investments.

The remainder of the chapter is organized as follows. Section 2 introduces a brief overview of green finance as a new model of sustainable finance, highlighting its main purposes and the relevance of environmental criteria. Section 3 describes the main green finance instruments that promote the financing of projects that are positive for the environment and the economy. Section 4 presents the role of green finance in encouraging companies to engage in green innovation. Section 5 examines the implications and challenges of green finance as a powerful tool for redirecting financial flows towards sustainable investments and green business innovations. Section 6 provides some concluding remarks.

2 From Sustainable Finance to Green Finance: Background

The growing awareness of environmental, social, and economic risks has stimulated continuous attention to the topic of sustainability. In the financial sector, the concept of sustainability implies the development of a financial intermediation model that incorporates alternative financial relationships, more responsible towards the economy and society, more suited to specific needs and more flexible and adaptable to the reference context. In this sense, the role of finance as an agent of constructive change has become fundamental, imposing the need to develop a sustainable finance model that includes activities that not only have the objective of generating a positive return, but which are capable of protecting the right of future generations to benefit from environmental, social and economic resources (Cunha et al. 2021; Kumar et al. 2022).

The European Commission defines sustainable finance as a dynamic process that takes ESG criteria into account when making financial and investment choices (Ren et al. 2023), highlighting the need to consider ESG criteria in investment strategies and financial decisions, which are the basis of sustainable development and can have a significant impact on long-term financial performance. Environmental criteria concern the evaluation of activities that reduce environmental impact, promote energy efficiency, sustainable management of natural resources and waste and pollution management. Social criteria concern the evaluation of activities that respect human rights, improve the quality of people's lives, promote workplace safety, diversity and inclusion, corporate social responsibility and social justice. Governance criteria concern the evaluation of the ethics and best practices of management activities through risk management policies, transparency, shareholder responsibility and regulatory compliance.

The growing importance of the role of sustainable finance which takes into account economic development together with social responsibility and above all respect for the environment has led to the creation of a new dedicated finance segment, namely green finance (Tao et al. 2022).

Specifically, while sustainable finance identifies a broader character of inclusiveness including social, environmental and economic aspects, green finance refers to financial activities whose proceeds are used for environmentally sustainable initiatives, with the sole objective of promoting a green economic transformation towards low-carbon, sustainable and inclusive pathways (United Nations Environment Programme 2016).

There is no real definition on the concept of green finance, but by analyzing the two terms separately, their meaning can be deduced. While finance calls for the transition of capital from one part of the world to another, ecology deals with problems related to the environment and the factors that regulate the presence and distribution of natural capital in different territories.

Green finance represents a "section" of finance that studies, on the one hand, the financial guidelines and strategies to be adopted to solve problems related to the environment (Dikau and Volz 2021), on the other, the financial interventions that can incentivize investments aimed at environmental protection (Wang and Zhi 2016).

According to the first perspective, green finance is based on the adoption of standards and guidelines for the implementation of responsible investment practices through which investors can choose to invest in activities that have a positive environmental impact, stimulating greater transparency and responsibility. In doing so, investors are able to identify environmental risks associated with certain investments and choose those that contribute to sustainability.

According to the second perspective, green finance identifies the presence of different types of financial instruments capable of supporting projects and activities that reduce environmental impact, promote energy efficiency and sustainable management of natural resources and contribute to the reduction of carbon emissions. carbon, thus promoting a more sustainable economy.

Green finance therefore becomes a strategic lever, even more so, an agent of constructive change for the development of a sustainable economy by directing financial resources towards environmentally friendly and socially responsible projects that bring positive environmental benefits and contribute to sustainable development (Gafoor et al. 2024; IFC 2016; Zhang and Zhao 2024). Indeed, in a world of global challenges such as climate change and resource depletion, capital mobilization is the crucial driving force to drive economies towards resilience and sustainability (Wang and Qu 2024).

Although there is no trace of the origin of the term green finance, it was widely used after the Earth Summit, held in Rio de Janeiro, Brazil, in 1992. At that meeting the topic of sustainable development was widely debated and it was reached to the conclusion that, to be sustainable, development must find a balance between three objectives—economic, social and environmental—in order to ensure today's well-being without compromising that of future generations.

Unfortunately, the results of the Conference were limited, due to the gap in interests between individual states. In fact, environmental and social problems can decrease with the improvement of the environmental and civil education of peoples and with the awareness that the solutions must be global and not sectoral, that is, they must concern the whole Earth and not just a part of it. Consequently, ecological sustainability is possible in a context of social development and economic growth and, then, the elimination of poverty becomes a crucial requirement for sustainable development. In fact, if today the most relevant environmental problems are global problems, there is no doubt, however, that poor populations are affected more than wealthy ones. Therefore, in the Declaration of the Summit on Social Development held in Copenhagen, three years after the Rio Conference, the Heads of State and Government committed themselves, starting then, to work to eliminate poverty in the world and create the conditions for sustainable development.

Various stakeholders have moved in this direction: from supranational and national governments, which operate through regulations and legislation on the matter, to government associations responsible for issuing ecolabels for products that respect the environment, to environmentalist associations that raise awareness public opinion on specific problems that have emerged, and to individual companies that use respect for the environment as a factor of qualitative distinction from competitors.

In particular, the European Commission has launched two important plans. The first is the Action plan on sustainable finance of 8 March 2018 aimed at regulating sustainable finance to build an economy with a reduced environmental impact (European Commission 2018). The second is the investment plan of the European Green Deal of 14 January 2020 aimed at achieving European environmental objectives and, in particular, achieving climate neutrality by 2050 (European Commission 2020).

Subsequently, the legislation on the Taxonomy for sustainable finance was introduced through European Union (EU) Regulation 2020/852 of 18 June 2020 (European Union 2020) which represents a practical guide for companies and investors to identify, on the basis of specific eco-sustainability criteria, the economic activities that could be considered sustainable.

The European Commission continued the path undertaken of environmental sustainability by strengthening the Taxonomy of the EU in the field of green which has allowed it to orient the horizons of various companies to be oriented towards greener investments. In particular, six environmental objectives have been identified that can allow an economic activity to be labeled as sustainable from an environmental point of view: climate change mitigation and adjustment; the sustainable use and protection of water and marine resources; the transition to a circular economy, considering waste prevention and growing the concentration of secondary raw resources; pollution prevention and control; the protection and renovation of ecosystems and biodiversity.

In this direction, the European Central Bank has also begun to require banking institutions to use the Taxonomy to grant concessions and incentives for financing with eco-sustainable purposes. Indeed, the financial sector, increasingly oriented towards green finance, can play a driving role in investments dedicated to environmental sustainability for the transition towards a neutral economy from the point of view of carbon emissions.

3 Green Finance: A New Model of Sustainability Finance

The market for green financial instruments has expanded rapidly in recent years, with investors increasingly interested in financing environmentally sustainable projects and aware of the limits of public funds and traditional financial instruments which have proven—in some cases—inadequate to satisfy environmental protection, curbing climate change, and to promote a new green business structure. In particular, the availability of appropriate financial instruments is observed for the support of economic activities aimed at environmental improvement, climate change and resource conservation and efficient use (Babic 2024; Sachs et al. 2019; Srivastava et al. 2022). The main green financial instruments can be found in the credit market, such as green loans, energy efficient mortgages, and carbon credits, and in the capital market such as green bonds and green funds.

3.1 *Green Credit Market*

In the credit market, some types of loans that can finance environmental sustainability are green loans, energy efficient mortgages, and carbon credits.

3.1.1 Green Loans

Green loans are a type of loan provided by financial institutions and banks to finance specific projects with environmental purposes. These projects can concern different areas of application, for example, renewable energy, waste management, energy efficiency, the reduction of greenhouse gas emissions, sustainable agriculture, clean mobility, and green real estate (Giraudet et al. 2021; Neagu et al. 2024). Green loans are characterized by a double advantage: on the one hand, they are characterized by some preferential conditions in terms of interest rate and preliminary costs compared to a traditional loan; on the other hand, they promote investment in activities that allow energy saving and environmental sustainability with a return over time in terms of reduction in expenses. Financial regulators should ensure a common definition and rules on credit institutions' green lending to also guarantee a level playing field for credit institutions and an adequate consumer protection for borrowers involved, and—at the same time—to promote the development and integrity of green loans (Ebeling 2022). In 2023, Loan Market Association (LMA 2023) published "Green Loans Principles" to provide guidelines to be used in the green loans market, clarifying when a loan can be classified as "green" and, whilst, allowing the loan product to maintain its flexibility. Four core components for alignment with the Green Loan Principles are defined. First, the proceeds of the green loans should be appropriately described in the finance documents, and environmental benefits should be clear to be assess and quantify by the borrower (Use of the proceeds). Then, the borrower of a green loan should disclose to its lenders the environmental sustainability objectives of the projects, the process by which the borrower determines how projects qualify as green projects, and additional information by which the borrower identifies and manages the social and environmental risks associated with the project (The process for project evaluation and selection). The borrower should track the net proceeds of the green loan by crediting them to a separate account to maintain transparency and ensure the integrity of the product (Management of proceeds). Finally, the reporting by borrowers should contain constantly updated information on the use of the proceeds until the loan expires and indicate the projects to which the proceeds of the green loans have been allocated (Reporting).

Looking at the latest data made available by the Climate Bond Initiative (Harrison 2023), 10.4 billion dollars of green loans are registered in 2022 with a cumulative total of 263 loan instruments, representing 2% of the market of which 70% of volume came from Asia–Pacific and Europe. The addition of 49 lending instruments contributed 19% to the cumulative total of 263. Green loans made up 2% of the market in 2022, and 70% of the volume came from Asia–Pacific and Europe combined (35% each).

3.1.2 Energy Efficient Mortgages

Energy Efficient Mortgages, on the other hand, are medium-long term loans aimed at energy efficiency projects, typically offered to private individuals by public agencies operating in the real estate market. The technical form of these mortgages is similar to that of traditional mortgages, but differs due to greater flexibility in the access parameters (loan/property value ratio or debt/income) and a series of incentive mechanisms linked to predetermined objectives of energy efficiency (Bell et al. 2023; Mathew et al. 2021). In 2018, as a function of the initiatives to promote energy efficiency in the EU, a group of leading international private banking institutions has joined the Energy Efficient Mortgage Action Plan (EeMAP) with the aim of giving impetus to the process of energy efficiency of the building stock and standardizing the investigation and disbursement procedures.

On 22 November 2022, the European Commission sent a Call for Advice to the European Banking Authority (2023) on green loans and mortgages, demanding its suggestion on the definition and potential supporting tools for these loans. 83 credit institutions from 27 countries participated in the survey, representing 52% of total assets in the banking sector in Europe. The survey shows that green loans represent 4.5% of the total loans of credit institutions, with a high percentage especially of loans to families (11%), compared to those to small and medium-sized enterprises (slightly over 2%), which are mainly granted for purchasing new properties.

3.1.3 Carbon Credits

Carbon credits are financial instruments that originate from companies' awareness of adopting low-carbon practices or technologies or reducing their greenhouse gas emissions to contribute to the fight against climate change (Demiralay et al. 2022; Swinkels and Yang 2023). The introduction of carbon credits was ratified in the Kyoto Protocol. The Paris Agreement validates the application of carbon credits to reduce greenhouse gas emissions and establishes provisions to further facilitate carbon credit markets.

Essentially, carbon credits allow companies to offset their greenhouse gas emissions by purchasing credits from other companies that have decreased their emissions. This reduction is verified by an independent accreditation body and converted into carbon credits. By purchasing carbon credits, companies can finance climate change mitigation projects and therefore contribute to reducing the number of emissions released in such a way as to avoid a net increase in atmospheric emissions globally.

Data on the voluntary carbon credit market highlights an increase from 2021, despite being totally based only on voluntariness and reputational objectives, thanks to the green policies of international agendas. In particular, the Report by Ecosystem Marketplace (2023) highlights that the value of the voluntary carbon market stood at just under 1.9 billion in 2022, reducing by only 11% compared to 2021 (2.2 billion). In detail, in the 2021–2022 period, overall commercial volumes decreased by 51%, going from 517 $MtCO_2e$ in 2021 to 254 $MtCO_2e$ in 2022 and recording a value of

49.2 $MtCO_2e$ in 2023. The average prices of credits on voluntary carbon markets, however, were higher than they have been in the last 15 years and increased in the 2021–2022 period by 82%, going from 4.04 dollars per ton in 2021 to 7.37 dollars per ton in 2022 and recording a value of 6.97 dollars per ton in 2023.

Project-linked credit data highlights that nature-based projects (forestry, land use and agriculture projects) and renewable energy projects continued to be the most popular categories for new projects in 2022 and 2023, demonstrating as market participants seek climate solutions across agriculture, forestry and other land uses that were previously underexplored.

3.2 *Green Capital Market*

In the capital market, some types of loans that can finance environmental sustainability are green bonds and green funds.

3.2.1 Green Bonds

Green bonds are debt securities issued by public or private entities, negotiable on the capital market, with the aim of raising capital to be allocated to initiatives with a low environmental impact, i.e. which incentivize the reduction of carbon emissions, improve efficiency energy, promote renewable energy, protect biodiversity (Zheng et al. 2023; Reboredo 2018). The fundamental characteristic of green bonds is that the issuer undertakes to use the proceeds exclusively to finance environmental or climate-friendly projects.

Green bonds are supported by an evaluation and certification framework capable of guaranteeing the transparency and correctness of information on the actual destination of funds to specific sustainable projects. In fact, this type of bond is researched by those investors who intend to finance a green initiative while also obtaining financial returns. Most of these bonds are certified by independent rating and certification agencies, such as the Climate Bonds Initiative or the Certification of Low Carbon Buildings. The International Capital Markets Association (ICMA 2021) published the guidelines, defined as Green Bond Principles, to allow investors to evaluate the green characteristics of a bond, guaranteeing the transparency and disclosure of the information necessary to also understand the assessment of the impact of their investments. Four core components for alignment with the Green Bond Principles are defined. First of all, the issuer identifies the activities in which to invest the proceeds deriving from the bond issue and the eco-sustainable objectives to be achieved (Use of proceeds). Subsequently, a broad-spectrum disclosure is necessary through which the criteria through which the issuer has qualified its project as green are made known (Process for project evaluation and selection). Then the issuer must prepare adequate internal processes to constantly keep track of the use of the proceeds collected and the percentage allocated to each green project linked to the issue (Management of proceeds). Finally, the issuer prepares a report of updated information regarding the use of the proceeds (Reporting).

Green bonds are distinguished in the following typologies:

- Standard green bond, that is a standard recourse-to-the-issuer debt obligation issued to collect funds with proceeds going to eligible green projects.
- Green revenue bond, that is a non-recourse-to-the-issuer debt obligation in which the credit exposure in the bond regards guaranteed cash flows of which the use of proceeds of the bond goes to green project.
- Green project bond, that is a project bond for single or multiple green projects for which the investor has direct exposure to the risk of the projects with or without potential recourse to the issuer, and its redemption depends on the cash flows of a single or multiple green projects.
- Green securitized bond, a bond collateralized by one or more specific green projects, including but not limited for example to asset backed securities and covered bonds. The first source of repayment is the cash flows of the assets.

The first green bond in the world is called Climate Awareness Bonds and it was issued in 2007 by the European Investment Bank for the financing of renewable energy and energy efficiency projects. In 2008, the World Bank also sent its first green bond. In the following years, the green bond market continued to expand with global green bond issuances rising from 14 billion of euros in 2014 to around 600 billion of euros in 2023. The EU also issued its first green bond in 2021, representing an example of the ambition of issuers and the interest of investors. The latest available data by Climate Bond Initiative (Harrison 2023) highlight that green bonds represent 65% of the total debt instruments linked to sustainability (16% social bonds, 13% sustainability bonds, 5% sustainability linked bonds, and 0.4% transition bonds) in 2023, registering a volume of 110.6 billion dollars (111.2 billion dollars in 2022). Regarding the geographical areas, the largest volume was recorded in Germany (117 billion dollars in 2023), while the United States recorded the largest number of deals (322 deals and 15.5 billion dollars).

Major private green bond issuers in 2023 include Renew Power with a 7.8 billion dollars bond issued to finance low-carbon energy; Baltic Power with 5.2 billion dollars bond to finance wind project; Volkswagen with 12.6 billion dollars bond to support electric vehicles.

3.2.2 Green Funds

Green funds fall into the broad category of socially responsible investments, which integrate ESG factors into investment decisions, but which have focused on environmental issues with the intensification of the climate crisis. A green fund is a mutual fund that invests in projects and companies engaged in activities that support the environment, such as reducing carbon emissions and promoting clean energy, conserving biodiversity, and managing natural resources. The investment strategies of a green fund involve the selection of companies that intend to support the environment, in compliance with precise screening criteria (Capotă et al. 2022; Gonçalves et al. 2021; Ibikunle and Steffen 2017).

The main objectives of green funds are:

- Promoting investments in environmentally responsive projects and technologies by focusing on sectors such as renewable energy, clean transport, energy efficiency, waste management, sustainable agriculture and conservation.
- Facilitating the transition towards a sustainable and low-carbon economy.
- Producing a financial return and, at the same time, generate positive environmental and social results for investors and the environment.

Specifically, some green funds simply build a portfolio of companies that do not rely on fossil fuels, deforestation or other unsustainable business activities. Other funds, however, actively seek to support companies engaged in research into new energies, sustainable materials or other technologies with environmental benefits (Chi et al. 2023).

Regarding risk management strategies, green funds exploit the advantages of diversification by choosing to include unrelated projects in their portfolio and continuously monitoring them to identify and address possible risks.

Within green funds, following the definition of new carbon risk measurement parameters by Morningstar in April 2018, low carbon mutual investment funds are also developing. In particular, fund managers have actively reduced their exposure to companies with high carbon risk scores (Ceccarelli et al. 2024).

Among the types of green funds there are:

- Low carbon funds, investing in companies with low carbon intensity compared to their benchmark. They offer multi-sector exposure and respond to the needs of those who want to reduce or completely eliminate climate risks from their portfolio.
- Climate transition funds, which select companies that consider climate change in their business strategy and that are therefore better prepared for the transition to a low-carbon economy. They are suitable for investors seeking to balance risk mitigation with the benefits of green transformation.
- Climate solution funds, who invest in companies that provide products and services that contribute to the transition to a low-carbon economy. They offer more concentrated exposure and are suitable for risk-averse investors who see climate change as an opportunity to generate value.
- Green bond funds, who invest in bonds issued to finance the transition to the green economy. By choosing these instruments, investors must evaluate the environmental solutions that are financed by green bonds.
- Clean energy and technology funds, who invest in companies that contribute to the transition towards clean energy, such as from renewable sources, the improvement of distribution and storage, etc. They are more specialized from a sector point of view and investors must be aware of the volatility that can characterize the sector.

According to the latest data available from Morningstar (Bioy 2023), Europe holds the largest number of green funds with 84% of total assets, followed by China with 8% and the United States (US) with 6%. The biggest investment is in climate transition funds and climate solutions funds, as investors look to invest in companies that consider climate change in their business strategy.

European investors tend to favor decarbonisation strategies and funds that focus on both risk and opportunity. Indeed, climate transition funds represent 45% of the assets of European climate funds, followed by climate solutions funds (19%) and clean energy and technology funds (9%). In contrast, US and Chinese investors prefer strategies focused on climate-related opportunities with climate solutions funds (31%) and clean energy and technology funds (46%) dominating total climate fund assets. However, US investors prefer clean energy and technology funds, while Chinese investors prefer both strategies equally.

4 Green Innovation in the Firms: The Role of Green Finance

The green finance market aims to incentivize the transfer of fund flows to economic entities engaged in environmental and eco-sustainable projects and activities. Expanding the reach of green finance is essential to facilitate the transition to a greener economic development model and achieve timely targets, while respecting the environmental dimension of the United Nations Sustainable Development Goals (Chen and Ma 2021).

In this context, an important role is played by firms which, as an integral component of the socio-economic fabric, play a dual role both as main contributors to polluting emissions and as key agents in pollution control (Ding et al. 2023). Effectively, the European Green Deal recognizes that "the private sector will be key to financing the green transition" and that "long-term signals are needed to direct financial and capital flows to green investment and to avoid stranded assets" (European Commission 2019, p. 16).

Therefore, a transition to a progressively greener future, characterized by the reduction of greenhouse gas emissions and the recovery of environmental degradation, requires a different economy. The latter is characterized by production processes and technologies that are more respectful of the environment and by a different conception of well-being, associated with new criteria. Through these criteria, firms can also evaluate the added value that they themselves produce based on the entire range of wealth and not just the flow of revenues and the quantity of machines and infrastructure accumulated.

This green innovation process describes technical advances that can improve the environment by reducing pollution and energy consumption (Usman and Hammar 2021). Green innovations are understood as "the introduction of any new or significantly improved product (good or service), process, organizational change or marketing solution that reduces the use of natural resources and decreases the release of harmful substances across the whole life-cycle" (EIO 2012, p. 8).

Like any investment, green innovations impact cost configurations and revenues with consequences on the economic and financial dynamics of firms (Ajibike et al. 2021). However, compared to traditional innovation, green innovation generates particular effects for companies in terms of corporate governance, production processes and cost–benefit trade-offs.

First of all, green innovation by companies requires a constant process of improving the conditions of use of environmental and human resources. This has immediate implications for corporate governance, as management and operational units must redefine much of the management and organizational methods with the adoption of eco-sustainable technologies and production processes. All aspects that introduce elements of uncertainty into the governance of the firm, increase complexity and complicate the decision-making process.

In most cases, green innovation presupposes changes in consolidated production processes, also requiring the availability of adequate tangible and intangible resources.

At the level of cost–benefit trade-offs, the adoption of eco-innovations can lead to high sacrifices in terms of wealth that are not compensated by similar immediate benefits. For example, investments in innovations that reduce greenhouse gas emissions or increase workplace safety lead to additional costs (investment and current management); these costs, by increasing production costs, penalize adopting companies compared to equivalent non-adopters, at least in the initial phases. It is therefore possible that the socially desirable outcome of less pollution associated with corporate activities may be incompatible with the objective of profit maximization (Horbach 2014).

Nonetheless, from a business point of view, there are multiple incentives towards the adoption of eco-innovations. These stimuli are linked to a plurality of both positive and negative elements. As regards the positive elements, the potential benefits of green investments concern:

- strengthening of the image towards customers, with consequent increase in customer satisfaction and legitimization by stakeholders of the initiatives launched; greater market shares despite the increase in sales prices of the product/service;
- increase in competitive advantage over competitors due to better environmental performance and eco-innovation strategies. This competitive advantage translates into indirect economic benefits of future valorisation which relate to the other dimensions of corporate development, i.e. social, environmental and competitive;
- development of new markets for firms mainly dedicated to the production of environmental goods and services (green core business) intended for other firms. In this context, for firms, which aspire to be market leaders or to position themselves in the high end of the market, sustainability needs are considered first and foremost a medium-long term investment, rather than a mere cost.

Instead, the negative elements refer above all to the need to adapt to competition and comply with regulations also to avoid possible sanctions. In particular, a greater push towards green innovation is plausible when companies are subject to strict environmental policy regulations and strong incentives (He et al. 2018).

In this context, finding financial resources to implement green innovation initiatives through traditional financing channels is one of the main problems that firms have had to face (Zameer and Yasmeen 2022). Indeed, until some time ago, firms that wanted to be sustainable had to bear the costs of limiting environmental impact. Today the sustainability interests of society and those of companies are aligned. Therefore, firms that invest in green innovation have access to better financing conditions and better market opportunities.

All this is made possible by green finance which has a positive impact on the environment and promotes sustainable development. The green finance market provides financial support to projects which, as mentioned, are difficult to finance within the traditional financial system. Green finance products create a virtuous circle whereby in providing capital to companies they incentivize the research and development of green investments, which are the key element that influences green innovation in companies (Flammer 2021).

Specifically, with green loans, banks can provide preferential interest rate loans to businesses to support their green investments. Some studies have shown how a reduction in financing costs affects the reduction of toxic emissions by stimulating investments in emission reduction activities (Goetz 2019). This does nothing but push polluting companies to review their production capacity, which is often obsolete, and encourages them to improve and innovate production technologies, thus increasing the level of green innovation. To obtain green loans, firms must comply with what is highlighted in the Green Loans Principles. First of all, it is necessary to provide the banks with the contractual documentation when structuring the loan. This documentation must contain an adequate description of the sustainable activities for which the financing is intended, the definition of the eco-sustainable objectives to be achieved and the criteria for quantifying the project as green. Secondly, firms must guarantee the traceability of the sums paid, setting up dedicated accounting and adequate internal processes. The firms also promptly make available updated information on the use of the sums disbursed, their destination and the development of the projects.

With carbon credits, firms can balance their carbon dioxide emissions, reducing their impact on climate change, or even offsetting more emissions than they have produced (Chen and Chen 2024). Carbon credits are important for firms especially in the context of sustainable management and environmental responsibility. In particular, adopting sustainable practices that lead to a reduction in carbon emissions can lead to significant savings in energy and operational costs. Firms can earn money by selling carbon credits generated by emissions reduction projects outside the firms themselves. Additionally, carbon credits can be traded on international markets, allowing firms to participate in global initiatives to reduce emissions. This can facilitate access to new markets and business opportunities. Several studies on carbon credit markets have focused attention on the need to improve corporate climate disclosure so that such disclosure does not compromise, rather than contribute to, global climate mitigation (Trouwloon et al. 2023).

Green bonds allow firms to raise financial resources in the capital markets to be allocated to activities related to climate change and, more generally, environmental performance. Although, from a technical point of view, green bonds have characteristics that are very similar to standard ones, some studies show that green bonds have lower financing costs, due to investors' environmental preferences and third-party certification (Sun et al. 2022; Zerbib 2019). Some studies highlight that the issuance of green bonds can improve ESG performance, while other studies show that the stock market responds positively to the issuance of green bonds (Gilchrist et al. 2021). As with green loans, when firms issue green bonds they must take green bond principles into account. These principles aim to strengthen credibility by proposing guidelines aimed at encouraging practices of transparency, disclosure and transmission of information (reporting). These principles were developed with the aim of providing all the information necessary to increase the capital allocated to environmental sustainability projects, and to support issuers during the transition of their business model towards greater environmental sustainability. These principles also provide useful tools for the ex post evaluation of the achievement of environmental sustainability objectives, and to improve the traceability of the activities carried out by the issuer following financial collection.

With green funds, socially and environmentally conscious business practices are promoted. Companies committed to respecting the environment, alternative energy and sustainable living are included in investment portfolios that attract different types of investors (Jia et al. 2023). All this guarantees new financial resources for businesses. Furthermore, by limiting the movement of capital towards firms that pollute and generate high energy consumption, green funds push firms to behave in a sustainable way by respecting the environment and encouraging green technological research and development. Some studies highlight how these funds not only produce a financial return but allow the creation of positive externalities for society and the environment (Allevi et al. 2019).

In essence, green finance improves the green innovation of firms not only because it finances environmental productivity, reducing polluting emissions and encouraging clean production (Zhang et al. 2019), but also because it has made financial operators more aware of the importance of environmental risk management (Wang et al. 2022). In this way, green finance, through financial instruments dedicated to environmental sustainability, is able, on the one hand, to attract investors with different risk preferences to invest, and on the other, to inspire green innovation by increasing environmental investments of firms.

5 Implications and Challenges

Green finance is based on the assumption that the operators of the financial and economic system use, in their financing choices, on the one hand, and investment choices, on the other, reference parameters that take into account the ability to guarantee the achievement of a level of environmental sustainability. The modification of behavior in a greener sense should lead to financing, more generally, all those entrepreneurial activities that produce an environmental and social benefit in an area, thus leading to greater prevention of environmental risks and the development of a sustainable economy.

Firms are no longer considered only in relation to their ability to produce a competitive advantage, but also and above all for the environmental and social value they are able to create (Al-Issa et al. 2022). Environmental innovation becomes an incentive for firms: their environmental and social policies rewarded through full recognition of the economic value generated. The inspiring idea is that reputation (or being publicly recognized as green firms, committed to respect for human rights and attentive to environmental issues) represents an important resource for its success. The firm that "behaves correctly" seems to generate greater value than the one that does not consider the principles of sustainable development. All this creates a virtuous cycle that leads to the financing of the most deserving firms by investors and to overall economic growth of society. For this reason, it is important to review the orientation of financial system operators which, through a range of products dedicated to the green theme, promote the selection of activities and projects that respect environmental criteria (Rahman et al. 2022).

Green finance pays particular attention to three relevant aspects:

- social performance, whereby every green finance intervention aims to have positive effects not only for the environment but also for the community and the territory. Just as traditional finance is aimed at maximizing individual profits, green finance is aimed at maximizing benefits for society;
- the transparency that must concern the activities and projects involved from all points of view to allow investors to choose to invest in activities with an environmental impact. After all, large firms have the obligation of non-financial reporting (Directive 2014/95/EU) relating to the use of renewable resources, the emission of pollutants, the impact of their activity on health and safety;
- integrity in the financial sector both in terms of autonomy from any form of conditioning, and in terms of correct management and compliance with regulations, in order to guarantee that investments are always oriented towards environmental and social objectives.

Through green finance it is possible to develop a sustainable economy, too. Indeed, the promotion of sustainable development is the consequence of the reacquisition of a mission of the economic-financial system focused on operational paradigms that intend to produce social well-being in compliance with environmental principles.

In this context, the construction of a self-sustainable economy, which does not neglect the environment, efficiency and society in favor of projects and financial flows, must be a priority (Yu et al. 2022). Indeed, the sustainability of development depends on the relationship between economic, social and environmental dimensions of the effects produced by every human activity. It is a dynamic equilibrium, the definition of which is linked to the progress of scientific and technical knowledge and to the construction of consensus on the best solutions to be adopted in order to guarantee access to resources that meets the criteria of fairness of the social community.

Implementing green finance is not without challenges. First of all, it is not easy to measure the environmental impact of investments and data on the performance of environmentally sustainable projects is often limited. In this sense, the assessment of risks associated with climate and the environment is also complex (Marczis et al. 2023). Furthermore, there is a lack of standardization and transparency in the market, making it difficult for investors to assess the environmental impact of their investments. For example, greater transparency of climate-related risks can facilitate a more adequate assessment of risks and capital allocation. Indeed, a taxonomy of green assets can facilitate financial institutions and investors to efficiently allocate capital. The presence of minimum standards that allow green finance to be defined in an increasingly transparent way and rules for disclosure are essential for redirecting capital flows towards green and sustainable investments, as well as for the analysis and benchmarking of the market and risks, also avoiding "greenwashing". In particular, greenwashing behavior is adopted when raising funds through green financial products without investing in green projects and falsely disclosing environmental information (Liu et al. 2024). The risk of exaggerated or misleading sustainability claims must be managed through an ongoing commitment to transparency and communication. Regulators and financial institutions should develop a regulatory framework that promotes truly sustainable practices while preventing misleading ones. The EU Action Plan for Sustainable Finance and the European Green Deal are just some examples of how regulation is directing capital towards more sustainable and transparent investments.

Finally, it is increasingly necessary to strengthen an environmental culture and awareness that leads to the adoption of tools that minimize environmental impacts or that can even enrich the ecosystem in which they are used. Indeed, green finance can not only propose new criteria for sustainable development, but can also stimulate the development of green projects and green technologies, allowing growth of the economic system while respecting the environment.

6 Conclusions

In recent years, the need to build the development of the global economy on sustainable bases has increasingly emerged, aiming for eco-sustainability and the green economy. In particular, it is important to reduce the waste of resources by concentrating investments in the renewable energy, climate, healthcare and water sectors. In this direction, green finance enhances the protection and respect of the environment and constitutes a valid complementary area to those typical of traditional finance, contributing, at the same time, to the production of wealth and the growth of economies.

The topic of green finance has generated numerous interventions aimed at safeguarding the environment by paying greater attention to the protection of "natural capital". A first milestone was reached in December 2009 when, in Copenhagen, useful rules were established to control, for example, the excessive emission of greenhouse gases and to decide how to deal with the impacts of climate change. Subsequently, greater awareness of green issues led to the implementation of financial initiatives aimed at the use of energy produced from renewable sources, the protection of forests and the introduction of adaptation measures useful for reducing environmental risks and damage. On the one hand, financial interventions should continue to enhance green finance by promoting the use of financial instruments dedicated to environmental issues. This implies the need for the financial sector to pay—at the same time—attention to the evaluation and monitoring of the environmental impact of the financed activities and to request a certification capable of guaranteeing the transparency and correctness of the information on the actual destination of the funds to specific sustainable projects. On the other hand, the growth of green finance has urged firms to include green development in the firm's strategic planning and to direct their investments towards the fight against pollution, the development of new waste technologies, the sustainable management of resources, as well as alternative energy and energy efficiency. In this sense, corporate green investments represent the beginning of an important transition towards a paradigm of sustainable economic development, acting as a basis for green transformation and innovation.

Therefore, it will be necessary to increase the promotion and recognition of financial institutions and green enterprises, support green consumption by the public, and create a good atmosphere to jointly build an ecological and civilized environment. In this direction, we should think for the future of a path towards alternative global policies that allow greater awareness and prevention of traditional resources, ensuring the population a sustainable future. The uncertainties on future evolution are reinforcing environmental concerns and efforts of EU countries, and they are consequently prompting them to improve their activities to diversify and expand ESG criteria. In this context, all policymakers and financial and economic institutions should take steps to support ESG criteria. Furthermore, commercial banks participate in green finance in the context of global environmental protection and sustainability trends. The role of European banks and the challenges that they face providing sustainability finance are important, although several are the difficulty of banks to find profitable business models that both work for the short- and long-term, respecting the ESG criteria.

Competing Interests The authors have no conflicts of interest to declare that are relevant to the content of this chapter.

References

Ajibike WA, Adeleke AQ, Mohamada F, Bamgbade JA, Nawi MNM, Moshooda TD (2021) An evaluation of environmental sustainability performance via attitudes, social responsibility, and culture: a mediated analysis. Environ Challenges 4. https://doi.org/10.1016/j.envc.2021.100161

Al-Issa N, Khaki AR, Jreisat A, Al-Mohamad S, Fahl D, Limani E (2022) Impact of environmental, social, governance, and corporate social responsibility factors on firm's marketing expenses and firm value: a panel study of US companies. Cogn Bus Manag 9(1). https://doi.org/10.1080/23311975.2022.2135214

Allevi E, Basso A, Bonenti F, Oggioni G, Riccardi R (2019) Measuring the environmental performance of green SRI funds: a DEA approach. Energy Econ 79(C):32–44. https://doi.org/10.1016/j.eneco.2018.07.023

Babic M (2024) Green finance in the global energy transition: actors, instruments, and politics. Energy Res Soc Sci 111. https://doi.org/10.1016/j.erss.2024.103482

Bell J, Battisti G, Guin B (2023) The greening of lending: evidence from banks' pricing of energy efficiency before climate-related regulation. Econ Lett 230. https://doi.org/10.1016/j.econlet.2023.111212

Bioy H (2023) Il patrimonio dei fondi climatici supera i 500 miliardi di dollari. *Morningstar*, Oct 9

Capotă LD, Giuzio M, Kapadia S, Salakhova D (2022) Are ethical and green investment funds more resilient? Working Paper Series, European Central Bank, Frankfurt, 2747, February

Ceccarelli M, Ramelli S, Wagner AF (2024) Low carbon mutual funds. Rev Fin 28(1):45–74. https://doi.org/10.1093/rof/rfad015

Chen W, Chen Z (2024) Production and carbon emission reduction decisions of remanufacturing firms with low-carbon credit financing under uncertain demand. Manag Decis Econ. https://doi.org/10.1002/mde.4182

Chen Y, Ma Y (2021) Does green investment improve energy firm performance? Energy Policy 153:112252. https://doi.org/10.1016/j.enpol.2021.112252

Chi Y, Hu N, Lu D, Yang Y (2023) Green investment funds and corporate green innovation: from the logic of social value. Energy Econ 119. https://doi.org/10.1016/j.eneco.2023.106532

Cunha FAFDS, Meira E, Orsato RJ (2021) Sustainable finance and investment: review and research agenda. Bus Strateg Environ 30(8):3821–3838. https://doi.org/10.1002/bse.2842

Demiralay S, Gencer HG, Bayraci S (2022) Carbon credit futures as an emerging asset: hedging, diversification and downside risks. Energy Econ 113. https://doi.org/10.1016/j.eneco.2022.106196

Dikau S, Volz U (2021) Central bank mandates, sustainability objectives and the promotion of green finance. Ecol Econ 184. https://doi.org/10.1016/j.ecolecon.2021.107022

Ding Q, Huang J, Chen J (2023) Does digital finance matter for corporate green investment? Evidence from heavily polluting industries in China. Energy Econ 117. https://doi.org/10.1016/j.eneco.2022.106476

Ebeling A (2022) European investment bank loan appraisal, the EU climate bank? Int Econ 172:203–216. https://doi.org/10.1016/j.inteco.2022.09.003

Ecosystem Marketplace (2023) Paying for quality state of the voluntary carbon markets 2023, ecosystem marketplace insights report, Nov 28

EIO (2012) Methodological report. Eco-innovation observatory. Funded by the EC, DG, Environment. Brussels
European Banking Authority (2023) EBA report in response to the call for advice from the European Commission on Green Loans and Mortgages, Paris, December
European Commission (2020) The European green deal investment plan and just transition mechanism explained. Brussels, 14 Jan
European Commission (2019) Communication from the commission to the European Parliament, the European Council, the Council, the European Economic and Social Committee and the Committee of the Regions. The European Green Deal. Brussels, Brussels, 11 Dec
European Commission (2018). Communication from the commission to the European parliament, the European council, the council, the European central bank, the European economic and social committee and the committee of the regions action plan: financing sustainable growth. COM(2018) 97 final, Brussels, 8 Mar
European Union (2020) Regulation (Eu) 2020/852 of the European Parliament and of the Council of 18 June 2020 on the establishment of a framework to facilitate sustainable investment, and amending Regulation (EU) 2019/2088, 22 June
Flammer C (2021) Corporate green bonds. J Financ Econ 142(2):499–516. https://doi.org/10.1016/j.jfineco.2021.01.010
Gafoor CPA, Perumbalath S, Daimari P, Naheem KT (2024) Trends and patterns in green finance research: a bibliometric study. Innov Green Dev 3(2). https://doi.org/10.1016/j.igd.2023.100119
Gilchrist D, Yu J, Zhong R (2021) The limits of green finance: a survey of literature in the context of green bonds and green loans. Sustainability 13(2). https://doi.org/10.3390/su13020478
Giraudet LG, Petronevich A, Faucheux L (2021) Differentiated green loans. Energy Policy 149:111861. https://doi.org/10.1016/j.enpol.2020.111861
Goetz M (2019) Financing conditions and toxic emissions. SAFE Working Paper Series, 254. https://doi.org/10.2139/ssrn.3411137
Gonçalves T, Pimentel D, Gaio C (2021) Risk and performance of European green and conventional funds. Sustainability 13(8). https://doi.org/10.3390/su13084226
Harrison C (eds) (2023) Sustainable debt market summary. Q3 2023. Climate Bond Initiative, London, November
He F, Miao X, Wong CWY, Lee S (2018) Contemporary corporate eco-innovation research: a systematic review. J Clean Prod 174:502–526. https://doi.org/10.1016/j.jclepro.2017.10.314
Horbach J (2014) Do eco-innovations need specific regional characteristics? An econometric analysis for Germany. Rev Reg Res 34(1):23–38. https://doi.org/10.1007/s10037-013-0079-4
Ibikunle G, Steffen T (2017) European green mutual fund performance: a comparative analysis with their conventional and black peers. J Bus Ethics 145(2). https://doi.org/10.1007/s10551-015-2850-7
ICMA (2021) Green bond principles. Voluntary process guidelines for issuing green bonds, Paris, June
IFC (2016) Green finance: a bottom-up approach to track existing flows. Washington, D.C.
Jia X, Li B, Liu ZF, Sun C (2023) The green effects of fund market—analysis based on institutional investors' preference. Kybernetes 52(2):495–517. https://doi.org/10.1108/K-11-2021-1120
Kumar S, Sharma D, Rao S, Lim WM, Mangla SK (2022) Past, present, and future of sustainable finance: insights from big data analytics through machine learning of scholarly research. Ann Oper Res. https://doi.org/10.1007/s10479-021-04410-8
Liu C, Wang J, Ji Q, Zhang D (2024) To be green or not to be: how governmental regulation shapes financial institutions' greenwashing behaviors in green finance. Int Rev Fin Anal 93(3). https://doi.org/10.1016/j.irfa.2024.103225
LMA (2023) Green loan principles. loan syndications & trading association, New York, February
Marczis D, Mihálovits Z, Sebestyén G (2023) Sustainability and climate risk data: a new era for investment decision-making in the age of climate change. Cogn Sustain 2(2). https://doi.org/10.55343/cogsust.64

Mathew P, Issler P, Wallace N (2021) Should commercial mortgage lenders care about energy efficiency? Lessons from a pilot study. Energy Policy 150. https://doi.org/10.1016/j.enpol.2021.112137
Neagu F, Tatarici L, Dragu F, Stamate A (2024) Are green loans less risky? Micro-evidence from a European emerging economy. J Fin Stab 70(4). https://doi.org/10.1016/j.jfs.2023.101208
Rahman S, Moral IH, Hassan M, Hossain GS, Perveen R (2022) A systematic review of green finance in the banking industry: perspectives from a developing country. Green Fin 4(3):347–363. https://doi.org/10.3934/GF.2022017
Reboredo JC (2018) Green bond and financial markets: co-movement, diversification and price spillover effects. Energy Econ 74:38–50. https://doi.org/10.1016/j.eneco.2018.05.030
Ren YS, Ma CO, Chen XQ, Lei YT, Wang YR (2023) Sustainable finance and blockchain: a systematic review and research agenda. Res Int Bus Fin 64(12). https://doi.org/10.1016/j.ribaf.2022.101871
Sachs JD, Woo WT, Yoshino N, Taghizadeh-Hesary F (2019) Why is green finance important? ADBI Working Paper, 917. https://doi.org/10.2139/ssrn.3327149
Srivastava AK, Dharwal M, Sharma A (2022) Green financial initiatives for sustainable economic growth: a literature review. Mater Today Proc 49(8):3615–3618. https://doi.org/10.1016/j.matpr.2021.08.158
Sun Z, Feng J, Zhou R, Yu Y, Deng Y (2022) Can labeled green bonds reduce financing cost in China? Sustainability 14(20). https://doi.org/10.3390/su142013510
Swinkels L, Yang J (2023) Investing in carbon credits. J Alternative Invest 26(2):28–59. https://doi.org/10.3905/jai.2023.1.195
Tao H, Zhuang S, Xue R, Cao W, Tian J, Shan Y (2022) Environmental finance: an interdisciplinary review. Technol Forecast Soc Chang 179:1–17. https://doi.org/10.1016/j.techfore.2022.121639
Trouwloon D, Streck C, Chagas T, Martinus G (2023) Understanding the use of carbon credits by companies: a review of the defining elements of corporate climate claims. Glob Challenges 7(4). https://doi.org/10.1002/gch2.202200158
United Nations Environment Programme (2016) Definitions and concepts: background note. Inquiry Working Paper, 16/13, Geneva, Switzerland
Usman M, Hammar N (2021) Dynamic relationship between technological innovations, financial development, renewable energy, and ecological footprint: fresh insights based on the STIRPAT model for Asia pacific economic cooperation countries. Environ Sci Pollut Res 28(12):1–18. https://doi.org/10.1007/s11356-020-11640-z
Wang L, Cheng Y, Wang Z (2022) Risk management in sustainable supply chain: a knowledge map towards intellectual structure, logic diagram, and conceptual model. Environ Sci Pollut Res 29(44):1–27. https://doi.org/10.1007/s11356-022-22255-x
Wang W, Qu Z (2024) Sustainable finance for promoting prosperity and recovery: the impact of mobilizing capital to expand natural resource efficiency. Resour Policy 91. https://doi.org/10.1016/j.resourpol.2024.104805
Wang Y, Zhi Q (2016) The role of green finance in environmental protection: two aspects of market mechanism and policies. Energy Procedia 104:311–316. https://doi.org/10.1016/j.egypro.2016.12.053
Yu M, Zhou Q, Cheok MY, Kubiczek J, Iqbal N (2022) Does green finance improve energy efficiency? New evidence from developing and developed economies. Econ Chang Restruct 55(1):485–509. https://doi.org/10.1007/s10644-021-09355-3
Zameer H, Yasmeen H (2022) Green innovation and environmental awareness driven green purchase intentions. Mark Intell Plan 40(5):624–638. https://doi.org/10.1108/MIP-12-2021-0457
Zerbib OD (2019) The effect of pro-environmental preferences on bond prices: evidence from green bonds. J Bank Finance 98:39–60. https://doi.org/10.1016/j.jbankfin.2018.10.012
Zhang D, Zhang Z, Managi S (2019) A bibliometric analysis on green finance: current status, development, and future directions. Fin Res Lett 29(8). https://doi.org/10.1016/j.frl.2019.02.003
Zhang T, Zhao F (2024) A study on the relationships among green finance, environmental pollution and economic development. Energ Strat Rev 51. https://doi.org/10.1016/j.esr.2023.101290

Zheng J, Jiang Y, Cui Y, Shen Y (2023) Green bond issuance and corporate ESG performance: steps toward green and low-carbon development. Res Int Bus Financ 66. https://doi.org/10.1016/j.ribaf.2023.102007

From Paper to Pixels for a World of Purpose: A Systematic Review of Sustainable Finance, Blockchain, and Digital Assets

Muhammad Azhar Khalil, Pathathai Sinliamthong, and Rashid Khalil

Abstract In a world grappling with pressing global challenges, the convergence of sustainable finance, blockchain technology, and digital assets emerges as a formidable driver in reshaping the financial landscape. This chapter undertakes a comprehensive exploration, navigating the terrain of sustainable finance, delving into its foundational principles, theories, and the integration of blockchain and digital assets within its framework. It underscores the transformative potential of blockchain technology, which enhances transparency, security, efficiency, and accessibility, thereby reinforcing the sustainable finance paradigm. The chapter offers an in-depth review of the field, supported by a thorough bibliometric analysis. The findings validate the dynamic nature of the field, characterized by mounting interest, interdisciplinary collaboration, and a pragmatic focus. The synergy of sustainable finance, blockchain technology, and digital assets presents a transformative trajectory towards a more responsible and sustainable financial future.

Keywords Sustainable finance · ESG · Blockchain · Digital assets · Sustainability · Responsible investing · FinTech

M. A. Khalil (✉) · P. Sinliamthong
Department of Finance and Risk Management, MSME Business School, Assumption University, Bangkok, Thailand
e-mail: mkhalil@msme.au.edu

P. Sinliamthong
e-mail: pathathaisnl@msme.au.edu

R. Khalil (✉)
Department of Banking and Finance, School of Business, Bahrain Polytechnic University, Isa Town, Bahrain
e-mail: rashid.khalil@polytechnic.bh

H. Baber and M. Fanea-Ivanovici (eds.), *Sustainable Financing—A Contemporary Guide for Green Finance, Crowdfunding and Digital Currencies*, World Sustainability Series, https://doi.org/10.1007/978-3-031-80969-9_5

1 Introduction

Sustainable finance represents a crucial response to global challenges, encompassing environmental, social, and governance (ESG) principles. The convergence of sustainable finance, blockchain technology, and digital assets holds the potential to revolutionize the financial sector (Ren et al. 2023). The confluence of sustainable finance, blockchain technology, and digital assets has emerged as a powerful force reshaping the financial landscape and addressing urgent global challenges. Sustainable finance, grounded in ESG principles, signifies a pivotal paradigm shift in investment approaches, aiming to redirect capital flows towards businesses, projects, and initiatives that promote sustainability, environmental preservation, and social wellbeing (Xiong and Dai 2023). Concurrently, blockchain technology has disrupted conventional financial models with its transparency, security, efficiency, and accessibility, laying the foundation for a more transparent and equitable financial ecosystem (Chung et al. 2023). However, it is the advent of digital assets, including cryptocurrencies, tokenized securities, and decentralized finance (DeFi) platforms, that adds a dynamic dimension to this transformative journey.

The escalating urgency of global challenges, such as climate change, resource scarcity, and social inequality, necessitates a fundamental reevaluation of financial practices. Sustainable finance, at its core, provides an innovative framework for addressing these concerns. It places emphasis on environmental sustainability, social responsibility, and robust governance practices in investment decision-making. As we navigate the complexities of ESG criteria, we realize that this approach is not merely an ethical or moral choice; it is an economic imperative. Investments aligned with sustainability and ESG principles are often linked to enhanced resilience, reduced risk, and more robust long-term financial performance (Khalil et al. 2022; Tahmid et al. 2022; Wang et al. 2023). This section sets the stage by elucidating the foundational concepts that guide sustainable finance and influence the integration of blockchain and digital assets within it.

Blockchain technology, often associated with cryptocurrencies like Bitcoin and Ethereum, has earned its place as a transformative tool underlying this confluence. The fundamental principles of transparency and immutability are central to blockchain's appeal (Qin et al. 2023). Transactions and records stored on a blockchain are tamper-resistant and verifiable, addressing the transparency aspect of ESG criteria (Schulz and Feist 2021). The decentralized and cryptographic nature of blockchain ensures a heightened level of security, a necessity for maintaining the integrity of sustainable finance initiatives (Agur et al. 2023). Moreover, blockchain's efficiency is derived from smart contracts, self-executing agreements that streamline financial processes, reducing administrative costs and risks (Natanelov et al. 2022). The technology's accessibility aspect is manifested through its potential to provide financial services to underserved populations, promoting financial inclusion, and aligning with the social pillar of ESG criteria. As we explore the multifaceted role of blockchain in sustainable finance, we unveil a dynamic force with the potential to instill accountability, efficiency, and accessibility in the financial sector.

The introduction of digital assets, which include cryptocurrencies like Bitcoin and tokenized securities, further revolutionizes sustainable finance. Cryptocurrencies, for instance, offer decentralized and borderless means of transferring value, reshaping traditional financial systems, and enabling more efficient and accessible financial solutions (Ghosh et al. 2020). In parallel, tokenized securities represent traditional financial assets transformed into digital tokens, enhancing accessibility and liquidity in sustainable investments. Decentralized Finance (DeFi) leverages blockchain and digital assets to create open and permissionless financial systems, enabling peer-to-peer lending, decentralized exchanges, and yield farming, thus advancing sustainable finance (Feng et al. 2022; Harwick and Caton 2022). As we explore these digital asset ecosystems, we witness a financial revolution that extends far beyond currency, reshaping how we invest, fundraise, and perceive the boundaries of financial systems.

The interplay between sustainable finance, blockchain, and digital assets presents an opportunity to shape a financial future that is not only economically viable but also socially and environmentally responsible. The transparency, security, efficiency, and accessibility offered by blockchain, and digital assets enhance the effectiveness and trustworthiness of sustainable finance initiatives. This systematic review endeavors to explore the intersection of sustainable finance, blockchain technology, and digital assets, delving into the fundamental concepts and theories underpinning each of these domains, from ESG investing principles to the transformative potential of blockchain and the extensive implications of digital assets. Adopting a systematic review methodology, we conduct a comprehensive synthesis of existing literature to provide insights into current trends, challenges, and opportunities at the nexus of sustainable finance, blockchain technology, and digital assets.

Research Questions: In pursuit of a deeper understanding of this interdisciplinary domain, our study is guided by the following research questions:

- What are the key principles and concepts underlying sustainable finance, blockchain technology, and digital assets?
- How do sustainable finance, blockchain technology, and digital assets intersect, and what are the potential synergies and challenges?
- What is the current state of research and practice in integrating sustainable finance, blockchain technology, and digital assets, and what are the implications for the financial sector?

By adopting a systematic review approach and addressing the research questions outlined above, this chapter aims to provide a comprehensive overview of the intersection of sustainable finance, blockchain technology, and digital assets, shedding light on the transformative potential of this interdisciplinary field.

The chapter delves into these concepts and theories, expanding on their nuances and exploring their interrelationship. In doing so, it provides a comprehensive overview of a transformative journey, one that holds the potential to steer the financial sector toward a greener and more socially responsible future. However, to fully grasp the profound implications of this intersection, it is imperative to address the challenges and limitations, including regulatory uncertainties, scalability issues, technical expertise requirements, and data privacy concerns that must be navigated. The chapter seeks to contribute to the ongoing discourse surrounding sustainable finance, blockchain, and digital assets and their role in shaping a more environmentally and socially responsible financial future. As this systematic review concludes, we gain a deeper understanding of how these challenges can be surmounted, laying the foundation for a financial future that embodies sustainability and responsibility at its core.

1.1 Current State of Research and Practice

The present state of research and practice at the confluence of sustainable finance, blockchain technology, and digital assets portrays a dynamic and evolving landscape. Ongoing initiatives, projects, and explorations by governments, international organizations, financial institutions, and innovative startups are actively reshaping the field. This section provides a deeper understanding of the current endeavors and trends, offering more convincing information regarding the state of research and practice in sustainable finance, blockchain, and digital assets as follows:

1.1.1 Government Initiatives

Governments worldwide have acknowledged the potential of blockchain and digital assets in advancing sustainability and are taking proactive measures to harness these technologies. For instance, the European Union (EU) has initiated the European Blockchain Partnership, a collaborative effort to explore blockchain applications across various sectors, including sustainable finance (Christodoulou et al. 2023). The EU's Blockchain Observatory and Forum continue to fund projects and research investigating blockchain's potential for environmental and societal sustainability (Ferrari 2018). This government support and investment in research and innovation underscore the strategic significance of these technologies in promoting sustainability.

1.1.2 International Organizations

Leading international organizations, such as the United Nations (UN) and the International Finance Corporation (IFC), have embarked on initiatives to integrate blockchain and digital assets into sustainable finance (de Villiers et al. 2021; Mulligan et al. 2023; Murimi et al. 2023). The UN's Sustainable Development Goals (SDGs) exemplify a global agenda that blockchain can support. Projects exploring blockchain for tracking supply chain sustainability of products and verifying the impact of investments are being championed by these organizations (Li et al. 2022). The Central Banks, through their digital transformation initiatives, are actively exploring how blockchain and digital assets can facilitate financial inclusion and support sustainable development (Copestake et al. 2023).

1.1.3 Financial Institutions

Prominent financial institutions, including banks, asset management firms, and investment platforms, are incorporating blockchain and digital assets into their operations. Leading investment firms are launching cryptocurrency funds, while banks are exploring blockchain for issuing green bonds and streamlining cross-border transactions. Notable examples include JPMorgan's development of its digital currency (Morgan 2022) and BlackRock's exploration of blockchain-based impact investment platforms (Ciriello 2021). These financial giants recognize the potential for blockchain and digital assets to transform sustainable finance.

1.1.4 Startups and Innovators

The startup ecosystem in sustainable finance, blockchain, and digital assets is vibrant and growing. Innovators are developing solutions that leverage blockchain's transparency and security to promote ESG compliance and impact measurement (Liu et al. 2023; Zhou and Liu 2023). Decentralized Finance (DeFi) projects are democratizing financial services, providing new avenues for impact investing and sustainability-linked loans (Chu et al. 2023). Impact-focused startups are utilizing tokenized securities to facilitate investments in renewable energy and conservation efforts.

1.1.5 Academic Research

Academic institutions are at the forefront of research in this field, with scholars investigating the theoretical underpinnings and practical applications of blockchain and digital assets in sustainable finance. Researchers are exploring topics such as the impact of cryptocurrencies on responsible financing (Anwer et al. 2023; Patel et al. 2023), the role of smart contracts in automating ESG compliance (Duchenne 2018), and the potential for tokenization to enhance liquidity in sustainable investments (Rachana Harish et al. 2023). Research centers and academic journals dedicated to this intersection are increasingly emerging, indicating a growing academic interest in this area.

1.1.6 Practical Applications

Beyond research, practical applications of blockchain and digital assets in sustainable finance are gaining traction. Companies are using blockchain to trace the environmental and social impact of products throughout their supply chains, enabling consumers to make informed and sustainable choices (Ciriello 2021; Natanelov et al. 2022). The tokenization of assets, from real estate (Kreppmeier et al. 2023) to renewable energy projects (Duchenne 2018), is becoming more prevalent, allowing fractional ownership and enhancing liquidity. Blockchain is also being used to streamline the issuance and management of green bonds (Christodoulou et al. 2023) and impact investments, promoting transparency in fund allocation.

The current state of research and practice at the intersection of sustainable finance, blockchain, and digital assets is marked by a convergence of efforts from governments, international organizations, financial institutions, startups, and academia. These diverse stakeholders are recognizing the transformative potential of blockchain and digital assets in advancing sustainability and reshaping the financial landscape. The practical applications and projects being undertaken are not confined to theoretical discussions but are actively contributing to a more sustainable and responsible financial future. The field is characterized by a continued commitment to innovation and collaboration, indicating a promising and influential path forward.

2 Background

2.1 Sustainable Finance: Key Concepts and Theories

Sustainable finance, guided by the principles of Environmental, Social, and Governance (ESG) criteria, is rooted in a multifaceted framework that seeks to redirect capital flows toward sustainable and responsible investments. It encompasses a range of concepts and theories such as the following:

2.1.1 ESG Criteria

ESG criteria serve as the bedrock of sustainable finance. Environmental criteria assess a company's impact on the natural world, emphasizing factors such as carbon emissions, water usage, and pollution (Jain et al. 2019). Social criteria evaluate a company's societal impact, including labor practices, employee relations, and community engagement (Arya and Zhang 2009). Governance criteria focus on the organization's structure and practices, including board diversity, executive compensation, and transparency (Dodd and Warner 1983). The integration of these criteria serves as a compass for sustainable investment decisions.

2.1.2 Impact Investing

Impact investing is a vital concept within sustainable finance. It goes beyond ESG criteria by explicitly targeting investments with the intention of generating measurable positive social or environmental impact alongside financial returns (Meuer et al. 2020). Impact investments can encompass areas such as renewable energy, affordable housing, or clean water initiatives, emphasizing the intent to create positive change.

2.1.3 Green Finance

Green finance refers to financial instruments and investment strategies that are explicitly designed to support environmentally sustainable projects and businesses (Khalil and Nimmanunta 2021). Green bonds, for example, are debt instruments whose proceeds are exclusively used to fund environmentally friendly projects. The development of green finance aligns with the environmental pillar of ESG principles.

2.1.4 Responsible Financing

Responsible financing is a broader concept that extends beyond ESG criteria and green finance. It encompasses the idea that financial institutions and investors have a responsibility to consider the broader societal and environmental implications of their actions (Jain et al. 2019). Responsible financing encourages a proactive approach to ensure financial activities do not contribute to negative environmental or social outcomes.

2.1.5 Sustainability Reporting

Sustainability reporting involves the disclosure of an organization's environmental and social performance. It is a key element of ESG investing as it provides transparency and accountability (Ting 2021). Organizations that adhere to sustainability reporting standards ensure that their ESG practices are clear and verifiable, aligning with ESG criteria.

2.2 Blockchain Technology and Its Integration in Sustainable Finance

Blockchain technology, characterized by its transparency, security, efficiency, and accessibility, has garnered attention for its potential to enhance sustainable finance. Blockchain's transparent and immutable ledger system ensures that transactions and records are tamper-resistant and verifiable. This aligns with the transparency aspect of ESG criteria. Researchers have emphasized blockchain's potential to eliminate fraud and corruption through its transparency (Liu et al. 2023).Blockchain's decentralized and cryptographic nature provides a high level of security, reducing the risk of fraud and unethical practices (Ghosh et al. 2020; Morgan 2022). The cryptographic techniques used in blockchain serve as a security measure. Blockchain's smart contracts, self-executing agreements embedded in the technology, can automate and streamline complex financial processes, reducing administrative costs and risks. Research has explored the efficiency gains that smart contracts can bring to financial transactions (Murimi et al. 2023).Blockchain can provide financial services to underserved populations, promoting financial inclusion and reducing inequality. The potential of blockchain to extend financial services to unbanked and underbanked populations has been noted (World Bank 2018).

2.3 *Digital Assets: An Emerging Dimension of Sustainable Finance*

Digital assets, including cryptocurrencies, tokenized securities, and decentralized finance (DeFi), have become pivotal components in sustainable finance:

Cryptocurrencies like Bitcoin and Ethereum offer decentralized and borderless means of transferring value. They have gained traction as a potential instrument for sustainable finance. Research has explored the potential of cryptocurrencies in supporting financial inclusion in developing countries, aligning with the social aspect of ESG criteria (Anwer et al. 2023). Tokenized securities represent traditional financial assets, such as stocks and bonds, transformed into digital tokens. They enhance accessibility and liquidity, making sustainable investments more flexible and attractive. Research has examined the potential of tokenized securities in improving the efficiency of financial markets (Ciriello 2021).

DeFi leverages blockchain and digital assets to create open and permissionless financial systems. Researchers have explored the transformative potential of DeFi in creating accessible and efficient financial services, particularly in underserved regions (Chu et al. 2023; Harwick and Caton 2022).

2.4 *The Confluence of Sustainable Finance, Blockchain, and Digital Assets: Opportunities and Challenges*

The intersection of sustainable finance, blockchain, and digital assets presents a multitude of opportunities for reshaping the financial sector, enhancing transparency, promoting responsible investments, and increasing accessibility. However, it is essential to acknowledge and address certain challenges such as regulatory uncertainty, scalability and energy consumption, required technical expertise, and data privacy.

The regulatory landscape for blockchain and cryptocurrencies is still evolving, creating uncertainty for sustainable finance initiatives. Research has explored the regulatory challenges and the need for a clear regulatory framework (Mbaidin et al. 2023). Some blockchain networks, especially early iterations like Bitcoin, have faced challenges related to scalability and energy consumption, which may not align with sustainability goals. Research has highlighted the energy consumption concerns of blockchain technology (Ma et al. 2023).

Implementing blockchain solutions requires technical expertise, which may be a barrier for many organizations, potentially limiting the adoption of these technologies. Research has explored the importance of technical knowledge in blockchain adoption (Chaganti et al. 2023). The transparency of blockchain may conflict with data privacy regulations, posing challenges in sensitive sectors like healthcare and finance. Research has discussed the privacy implications of blockchain technology (Qahtan et al. 2023).

3 Methodology

Conducting a systematic review on the intersection of sustainable finance, blockchain technology, and digital assets requires a rigorous data collection and analysis methodology. This section outlines the data sources, search strategy, inclusion criteria, and bibliometric analysis techniques used in this systematic review.

3.1 Data Sources and Search Strategy

The data collection process for this systematic review aimed to comprehensively capture relevant literature at the intersection of sustainable finance, blockchain technology, and digital assets. A combination of scholarly databases, academic publications, and grey literature sources were utilized to ensure a thorough exploration of the topic. The primary databases used for this study are presented in Table 5.1.

The search strategy employed a combination of keywords and phrases, including but not limited to blockchain technology, sustainability, sustainable development, sustainable finance, impact investing, ESG investing, digital assets, DeFi (decentralized finance), green bonds, artificial intelligence, cryptocurrencies, tokenized securities, responsible financing, smart contracts, security, supply chain management, and industry 4.0. The search strategy aimed to capture a comprehensive set of articles and documents that addressed the intersection of these domains.

Table 5.1 Databases used for searching relevant literature

Database	Description
PubMed	Searched for articles with relevance to sustainable finance, blockchain technology, and digital assets, with a focus on healthcare and medical applications
IEEE Xplore	Searched for articles related to blockchain technology and its applications, particularly in finance and healthcare
Google Scholar	Conducted broad searches for publications across various fields, with a focus on articles that explore the intersection of sustainable finance, blockchain, and digital assets
ScienceDirect	Searched for articles on sustainable finance, ESG criteria, and their integration with blockchain and digital assets
Web of science	Conducted searches for scholarly articles, conference papers, and proceedings related to sustainable finance, blockchain, and digital assets
Grey literature sources	Included reports, white papers, and documents from organizations such as the World Bank, IMF, and relevant industry associations
Search result: 580 articles	

3.2 *Inclusion and Exclusion Criteria*

The selection of articles and documents adhered to specific inclusion and exclusion criteria to ensure the relevance and quality of the literature reviewed. The criteria were refined to focus on materials directly addressing the intersection of sustainable finance, blockchain, and digital assets, thereby excluding irrelevant or outdated materials, as outlined in Table 5.2.

Table 5.2 Criteria for selecting articles

Inclusion criteria	Exclusion criteria
• Articles and documents published from 2000 to the present, reflecting recent developments in sustainable finance, blockchain, and digital assets • Research articles, conference papers, reports, and relevant academic literature • Materials addressing sustainable finance, ESG criteria, blockchain technology, and digital assets • Materials exploring the integration of blockchain and digital assets in sustainable finance	• Articles in languages other than English, given the primary language of this review • Materials that did not address the intersection of sustainable finance, blockchain, and digital assets • Outdated or irrelevant materials that did not contribute to the goals of this review

3.3 Bibliometric Analysis

To conduct a comprehensive bibliometric analysis, we use Bibliometrix R-package and its webapp namely VOSviewer, and Excel to explore the publication trends, authors and affiliations, citation analysis, keywords and co-occurrence analysis, journal and source analysis, and geographical distribution in order to comprehensively understand the research questions and objectives formulated in the introduction section. The analysis included several databases including Web of Science, Scopus, and PubMed. The parameters used for the bibliometric analysis are reported in Table 5.3.

By employing these bibliometric methods, this systematic review provides a comprehensive overview of the current state of research and practice in sustainable finance, blockchain, and digital assets and contributes to a deeper understanding of this dynamic and evolving field.

Table 5.3 Parameters used to perform bibliometric analysis

Parameters	Description/Research Question
Publication trends	An analysis of the publication trends over the years was carried out to identify the evolving interest and research activity in the intersection of sustainable finance, blockchain, and digital assets • How has research activity in the intersection of sustainable finance, blockchain, and digital assets evolved?
Authors and affiliations	The study examined prolific authors and affiliations to identify key contributors to the field and potential academic and industry leaders • Who are the key contributors and potential academic and industry leaders in this field?
Citation analysis	The review explored highly cited articles to identify influential works and their impact on the field • What are the most influential works and their impact on the intersection of sustainable finance, blockchain, and digital assets?
Keywords and co-occurrence analysis	The study identified prominent keywords and terms, including co-occurrence of terms, to understand the key themes and topics within the research literature • What are the key themes and topics within the research literature, as indicated by prominent keywords and co-occurrence analysis?
Journal and source analysis	The analysis included an examination of the journals and sources publishing the most relevant articles, highlighting respected and authoritative outlets in the field • Which journals and sources are publishing the most relevant articles in this field, and what are their respective impact factors?
Geographical distribution	The study considered the geographical distribution of authors and their affiliations to discern regional interests and trends in the intersection of sustainable finance, blockchain, and digital assets • What are the regional interests and trends in research related to sustainable finance, blockchain, and digital assets?

4 Results

To summarize the primary research directions within this field, this study presents statistical data on frequently occurring keywords and the number of published works. Figure 5.1 illustrates 15 keywords and concepts that are commonly found, highlighting prevalent research topics in the literature collection. Figure 5.1 notably underscores the significant role of blockchain. Furthermore, the rise of cutting-edge technologies such as artificial intelligence, and smart contracts indicates a trend in current blockchain research, where integration with other advanced technologies is sought due to the inherent limitations of blockchain technology by itself. Researchers aim to fortify and complement blockchain with additional technologies to foster sustainable development across various domains by synergizing and innovating with diverse technologies (Yadav et al. 2020).

In contrast, blockchain technology has disrupted regulations in multiple sectors, including market, finance, and supply chain, necessitating a reevaluation of production relationships and industry norms (Yadav and Singh 2020). This shift has ushered in the industrialization of the circular economy, green finance, and sustainable development. Blockchain is poised to reshape the supply chain system and redefine financial collaboration mechanisms, enhance asset optimization and scheduling, and boost resource circulation rates (Tseng et al. 2021).

Overall, the available evidence showcases the extensive and multidimensional research efforts carried out in the financial sector regarding blockchain technology. It has facilitated the convergence and harmonization of sustainable finance principles with the implementation of novel technologies. Furthermore, it has illustrated

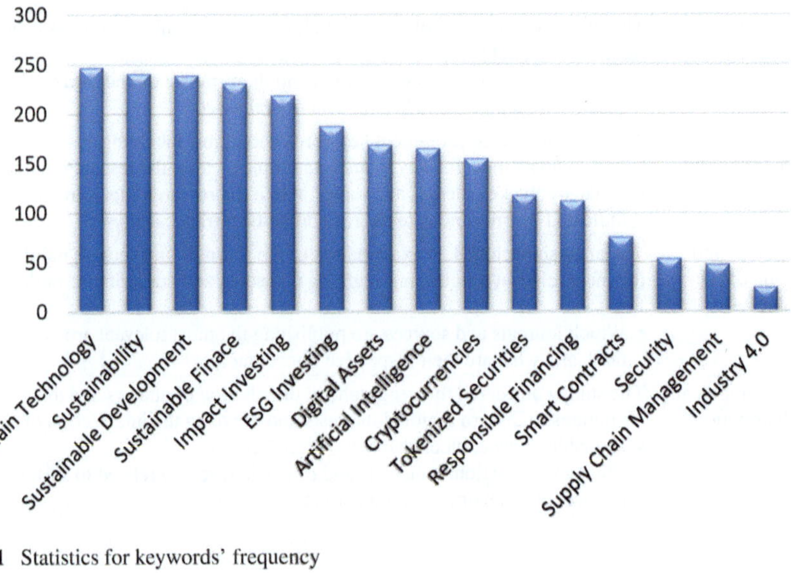

Fig. 5.1 Statistics for keywords' frequency

the remarkable versatility of blockchain technology, highlighting its adaptability to various industries.

The bibliometric analysis conducted in this systematic review reveals several key trends and insights regarding the intersection of sustainable finance, blockchain technology, and digital assets. Notably, there has been a significant increase in research activity related to these areas in recent years, aligning with previous research findings. Prominent authors and affiliations exhibit diversity, representing academia and industry, consistent with prior studies. Highly cited articles often focus on practical applications of blockchain and digital assets in sustainable finance, indicating a shift toward real-world implementation. Key themes and topics within the literature include sustainable finance, ESG criteria, blockchain technology, digital assets, and impact investing, with a growing emphasis on interdisciplinary explorations. Scholarly journals, academic conferences, and reports from international organizations serve as prominent sources of information, highlighting the multi-sectoral interest and collaborative nature of research in this domain. The geographical distribution of authors and affiliations reveals a global interest in sustainable finance, blockchain, and digital assets, consistent with previous research findings. Furthermore, the analysis revealed specific insights regarding the intersection of sustainable finance, blockchain technology, and digital assets, including the following:

4.1 Publication Trends

The analysis of publication trends demonstrates a significant increase in research activity related to sustainable finance, blockchain, and digital assets in recent years. This aligns with the growing recognition of the importance of these areas in addressing global sustainability challenges. The uptick in publications is in line with previous research (Mbaidin et al. 2023; Rachana Harish et al. 2023), indicating a continuous and intensifying interest in these interconnected domains.

4.2 Authors and Affiliations

Prominent authors and affiliations in this field exhibit diversity, representing both academia and industry. Notable contributions come from experts in finance, technology, and sustainability, reflecting the interdisciplinary nature of the research. This observation is consistent with previous studies (Ciriello 2021; Dal Mas et al. 2023; Liu et al. 2023) that have identified a wide array of contributors from various sectors.

4.3 Citation Analysis

Highly cited articles often focus on the practical applications of blockchain and digital assets in sustainable finance. These influential works tend to explore use cases, such as the impact of blockchain on supply chain sustainability and the role of cryptocurrencies in impact investing. The emphasis on practical applications suggests that the field is evolving from theoretical discussions to real-world implementation. This trend aligns with previous research (Agur et al. 2023; Feng et al. 2022; Schulz and Feist 2021), indicating that blockchain's impact on sustainability is not merely theoretical but has practical implications.

4.4 Keywords and Co-occurrence Analysis

Key themes and topics within the research literature include "sustainable finance," "ESG criteria," "blockchain technology," "digital assets," and "impact investing." The co-occurrence of these terms underscores the interconnectedness of these domains and their relevance to one another. Research on these topics has evolved from isolated discussions to interdisciplinary explorations (Ciriello 2021; Ren et al. 2023; Xiong and Dai 2023) that aim to identify synergies among sustainable finance, blockchain technology, and digital assets.

4.5 Journal and Source Analysis

Scholarly journals, academic conferences, and reports from international organizations are among the prominent sources of information in this field. The high-quality sources published on this topic provide a foundation of credible and authoritative research. The diversity of sources underlines the multi-sectoral interest and collaborative nature of research in this domain, which is consistent with the findings of previous research (Duchenne 2018).

4.6 Geographical Distribution

The geographical distribution of authors and affiliations reveals a global interest in sustainable finance, blockchain, and digital assets. Researchers, both in academia and industry, hail from diverse regions worldwide. While certain countries, particularly those with well-established financial and technological hubs, contribute significantly, the global distribution underscores the universality and cross-border relevance of the field. This finding resonates with previous research (Mulligan et al. 2023; Qin et al.

2023; Ren et al. 2023), which indicates that sustainable finance and blockchain are global phenomena.

The results of the bibliometric analysis demonstrate a growing body of research and practice at the intersection of sustainable finance, blockchain technology, and digital assets. The trends indicate that the field is maturing, with a focus on practical applications and interdisciplinary collaboration. The diverse range of authors and affiliations, the influence of highly cited works, and the global distribution of contributors all attest to the significance of these interconnected domains in reshaping the financial landscape towards a more sustainable and responsible future.

5 Conclusion and Implications

The confluence of sustainable finance, blockchain technology, and digital assets presents a transformative force reshaping the financial sector. The objectives outlined in the previous section aimed to explore the intersection of sustainable finance, blockchain technology, and digital assets, and understand their integration, potential benefits, challenges, and implications for the financial sector. The study in this chapter addresses these objectives by highlighting the transformative potential of these domains and their role in addressing pressing global challenges. The analysis confirms the hypotheses posited in the introduction, demonstrating a vibrant and evolving landscape characterized by increasing research activity, practical applications, and collaborations across sectors and regions. The findings validate the importance of sustainable finance, blockchain, and digital assets in reshaping the financial sector towards a more conscientious and sustainable future.

The rise of sustainable finance is a response to the pressing global challenges of our time, from climate change to social inequality. Its integration with blockchain technology aligns with the principles of ESG criteria, enhancing transparency, security, efficiency, and accessibility. The advent of digital assets, including cryptocurrencies, tokenized securities, and DeFi, expands the scope of sustainable finance, offering decentralized and innovative financial solutions.

The intersection of these domains presents numerous opportunities for reshaping the financial sector. These opportunities include enhancing ESG compliance, promoting responsible investments, increasing financial inclusion, and streamlining financial processes. However, there are also challenges to address, including regulatory uncertainties, scalability issues, technical expertise requirements, and data privacy concerns. The findings of this systematic review underline the vibrant and evolving nature of this field. Prominent actors, including governments, international organizations, financial institutions, startups, and academia, are actively contributing to its development. Collaborations across sectors and regions are resulting in a multi-faceted body of research that aims to address the complex challenges of our time.

As the financial landscape continues to evolve, sustainable finance, blockchain, and digital assets will play pivotal roles in shaping a more conscientious and sustainable future. Their transformative potential in addressing global challenges offers a path forward that is both economically viable and socially and environmentally responsible. This systematic review contributes to our understanding of the current state of research and practice in these domains and underscores their collective potential to influence the future of finance.

While our study provides valuable insights into the current state of research and practice in these domains, several directions for future research warrant exploration. These include addressing regulatory uncertainties, scalability issues, technical expertise requirements, and data privacy concerns. Additionally, further investigation into the practical implications of integrating blockchain technology with sustainable finance and digital assets is needed to inform strategic decision-making and policy formulation.

Competing Interests The authors have no conflicts of interest to declare that are relevant to the content of this chapter.

References

Agur I, Lavayssière X, Villegas Bauer G, Deodoro J, Martinez Peria S, Sandri D, Tourpe H (2023) Lessons from crypto assets for the design of energy efficient digital currencies. Ecol Econ 212:107888. https://doi.org/10.1016/J.ECOLECON.2023.107888

Anwer Z, Farid S, Khan A, Benlagha N (2023) Cryptocurrencies versus environmentally sustainable assets: does a perfect hedge exist? Int Rev Econ Financ 85:418–431. https://doi.org/10.1016/J.IREF.2023.02.005

Arya B, Zhang G (2009) Institutional reforms and investor reactions to CSR announcements: evidence from an emerging economy. J Manage Stud 46(7):1089–1112. https://doi.org/10.1111/J.1467-6486.2009.00836.X

Chaganti R, Bhushan B, Ravi V (2023) A survey on blockchain solutions in DDoS attacks mitigation: techniques, open challenges and future directions. Comput Commun 197:96–112. https://doi.org/10.1016/J.COMCOM.2022.10.026

Christodoulou P, Psillaki M, Sklias G, Chatzichristofis SA (2023) A blockchain-based framework for effective monitoring of EU green bonds. Financ Res Lett 58:104397. https://doi.org/10.1016/J.FRL.2023.104397

Chu J, Chan S, Zhang Y (2023) An analysis of the return–volume relationship in decentralised finance (DeFi). Int Rev Econ Financ 85:236–254. https://doi.org/10.1016/J.IREF.2023.01.006

Chung KHY, Li D, Adriaens P (2023) Technology-enabled financing of sustainable infrastructure: a case for blockchains and decentralized oracle networks. Technol Forecast Soc Chang 187:122258. https://doi.org/10.1016/J.TECHFORE.2022.122258

Ciriello RF (2021) Tokenized index funds: a blockchain-based concept and a multidisciplinary research framework. Int J Inf Manage 61:102400. https://doi.org/10.1016/J.IJINFOMGT.2021.102400

Copestake A, Furceri D, Gonzalez-Dominguez P (2023) Crypto market responses to digital asset policies. Econ Lett 222:110949. https://doi.org/10.1016/J.ECONLET.2022.110949

Dal Mas F, Massaro M, Ndou V, Raguseo E (2023) Blockchain technologies for sustainability in the agrifood sector: a literature review of academic research and business perspectives. Technol Forecast Soc Chang 187:122155. https://doi.org/10.1016/J.TECHFORE.2022.122155

de Villiers C, Kuruppu S, Dissanayake D (2021) A (new) role for business—promoting the United Nations' sustainable development goals through the internet-of-things and blockchain technology. J Bus Res 131:598–609. https://doi.org/10.1016/J.JBUSRES.2020.11.066

Dodd P, Warner JB (1983) On corporate governance: a study of proxy contests. J Financ Econ 11(1–4):401–438. https://doi.org/10.1016/0304-405X(83)90018-1

Duchenne J (2018) Blockchain and smart contracts: complementing climate finance, legislative frameworks, and renewable energy projects. Transform Clim Fin Green Invest Blockchains 303–317. https://doi.org/10.1016/B978-0-12-814447-3.00022-7

Feng S, Zhang R, Li G (2022) Environmental decentralization, digital finance and green technology innovation. Struct Chang Econ Dyn 61:70–83. https://doi.org/10.1016/J.STRUECO.2022.02.008

Ferrari V (2018) EU blockchain observatory and forum workshop on GDPR, data policy and compliance. SSRN Electron J. https://doi.org/10.2139/SSRN.3247494

Ghosh A, Gupta S, Dua A, Kumar N (2020) Security of cryptocurrencies in blockchain technology: state-of-art, challenges and future prospects. J Netw Comput Appl 163:102635. https://doi.org/10.1016/J.JNCA.2020.102635

Harwick C, Caton J (2022) What's holding back blockchain finance? On the possibility of decentralized autonomous finance. Q Rev Econ Finance 84:420–429. https://doi.org/10.1016/J.QREF.2020.09.006

Jain M, Sharma GD, Srivastava M (2019) Can sustainable investment yield better financial returns: a comparative study of ESG indices and MSCI indices. Risks 7(1):15. https://doi.org/10.3390/RISKS7010015

Khalil MA, Khalil R, Khalil MK (2022) Environmental, social and governance (ESG)—augmented investments in innovation and firms' value: a fixed-effects panel regression of Asian economies. China Fin Rev Int. https://doi.org/10.1108/CFRI-05-2022-0067

Khalil MA, Nimmanunta K (2023) Conventional versus green investments: advancing innovation for better financial and environmental prospects. J Sustain Fin Invest 13(3):1153–1180. https://doi.org/10.1080/20430795.2021.1952822

Kreppmeier J, Laschinger R, Steininger BI, Dorfleitner G (2023) Real estate security token offerings and the secondary market: driven by crypto hype or fundamentals? J Bank Finance 154:106940. https://doi.org/10.1016/J.JBANKFIN.2023.106940

Li Q, Ma M, Shi T, Zhu C (2022) Green investment in a sustainable supply chain: the role of blockchain and fairness. Transp Res Part E Logist Transp Rev 167:102908. https://doi.org/10.1016/J.TRE.2022.102908

Liu X, Yang Y, Jiang Y, Fu Y, Zhong RY, Li M, Huang GQ (2023) Data-driven ESG assessment for blockchain services: a comparative study in textiles and apparel industry. Resour Conserv Recycl 190:106837. https://doi.org/10.1016/J.RESCONREC.2022.106837

Ma C-Q, Lei Y-T, Ren Y-S, Chen X-Q, Wang Y-R, Narayan S (2023) Systematic analysis of the blockchain in the energy sector: trends, issues, and future directions. Telecommun Policy: 102677. https://doi.org/10.1016/J.TELPOL.2023.102677

Mbaidin HO, Alsmairat MAK, Al-Adaileh R (2023) Blockchain adoption for sustainable development in developing countries: challenges and opportunities in the banking sector. Int J Inf Manage Data Insights 3(2):100199. https://doi.org/10.1016/J.JJIMEI.2023.100199

Meuer J, Koelbel J, Hoffmann VH (2020) On the nature of corporate sustainability. Org Environ 33(3):319–341. https://doi.org/10.1177/1086026619850180

Morgan PJ (2022) Assessing the risks associated with green digital finance and policies for coping with them. Econ Law Inst Asia Pacific: 51–68. https://doi.org/10.1007/978-981-19-2662-4_3/COVER

Mulligan C, Morsfield S, Cheikosman E (2023) Blockchain for sustainability: a systematic literature review for policy impact. Telecommun Policy: 102676. https://doi.org/10.1016/J.TELPOL.2023.102676

Murimi R, Bell G, Rasheed AA, Beldona S (2023) Blockchains: a review and research agenda for international business. Res Int Bus Financ 66:102018. https://doi.org/10.1016/J.RIBAF.2023.102018

Natanelov V, Cao S, Foth M, Dulleck U (2022) Blockchain smart contracts for supply chain finance: Mapping the innovation potential in Australia-China beef supply chains. J Ind Inf Integr 30:100389. https://doi.org/10.1016/J.JII.2022.100389

Patel R, Kumar S, Bouri E, Iqbal N (2023) Spillovers between green and dirty cryptocurrencies and socially responsible investments around the war in Ukraine. Int Rev Econ Financ 87:143–162. https://doi.org/10.1016/J.IREF.2023.04.013

Qahtan S, Yatim K, Zulzalil H, Osman MH, Zaidan AA, Alsattar HA (2023) Review of healthcare industry 4.0 application-based blockchain in terms of security and privacy development attributes: comprehensive taxonomy, open issues and challenges and recommended solution. J Netw Comput Appl 209:103529. https://doi.org/10.1016/J.JNCA.2022.103529

Qin M, Zhang X, Li Y, Badarcea RM (2023) Blockchain market and green finance: the enablers of carbon neutrality in China. Energy Econ 118:106501. https://doi.org/10.1016/J.ENECO.2022.106501

Rachana Harish A, Wu W, Li M, Huang GQ (2023) Blockchain-enabled digital asset tokenization for crowdsensing in environmental, social, and governance disclosure. Comput Ind Eng 185:109664. https://doi.org/10.1016/J.CIE.2023.109664

Ren YS, Ma CQ, Chen XQ, Lei YT, Wang YR (2023) Sustainable finance and blockchain: a systematic review and research agenda. Res Int Bus Financ 64:101871. https://doi.org/10.1016/J.RIBAF.2022.101871

Schulz K, Feist M (2021) Leveraging blockchain technology for innovative climate finance under the green climate fund. Earth Syst Govern 7:100084. https://doi.org/10.1016/J.ESG.2020.100084

Tahmid T, Hoque MN, Said J, Saona P, Azad MAK (2022) Does ESG initiatives yield greater firm value and performance? New evidence from European firms. Cogn Bus Manage 9(1):1–16. https://doi.org/10.1080/23311975.2022.2144098

Ting PH (2021) Do large firms just talk corporate social responsibility? The evidence from CSR report disclosure. Financ Res Lett 38:101476. https://doi.org/10.1016/J.FRL.2020.101476

Tseng ML, Bui TD, Lim MK, Tsai FM, Tan RR (2021) Comparing world regional sustainable supply chain finance using big data analytics: a bibliometric analysis. Ind Manag Data Syst 121(3):657–700. https://doi.org/10.1108/IMDS-09-2020-0521/FULL/PDF

Wang J, Hu X, Zhong A (2023) Stock market reaction to mandatory ESG disclosure. Financ Res Lett 53:103402. https://doi.org/10.1016/J.FRL.2022.103402

World Bank (2018) Blockchain and emerging digital technologies for enhancing post-2020 climate markets. https://doi.org/10.1596/29499

Xiong Y, Dai L (2023) Does green finance investment impact on sustainable development: role of technological innovation and renewable energy. Renew Energy 214:342–349. https://doi.org/10.1016/J.RENENE.2023.06.002

Yadav G, Kumar A, Luthra S, Garza-Reyes JA, Kumar V, Batista L (2020) A framework to achieve sustainability in manufacturing organisations of developing economies using industry 4.0 technologies' enablers. Comput Industr 122:103280. https://doi.org/10.1016/J.COMPIND.2020.103280

Yadav S, Singh SP (2020) Blockchain critical success factors for sustainable supply chain. Resour Conserv Recycl 152:104505. https://doi.org/10.1016/J.RESCONREC.2019.104505

Zhou H, Liu J (2023) Digitalization of the economy and resource efficiency for meeting the ESG goals. Resour Policy 86:104199. https://doi.org/10.1016/J.RESOURPOL.2023.104199

Green Finance Strategies: Driving Sustainable Business Models Through Innovative Financing Solutions

Itishree Behera and Pragyan Nanda

Abstract At the cutting edge of driving sustainable business models, green finance offers innovative financing solutions that support environmentally responsible practices. This chapter comprehensively investigates green finance strategies and their impact on sustainable business development. It begins with an overview of the theoretical foundations of green finance, emphasizing its significance in sustainable development. Various innovative financing solutions, including green bonds, green loans, and impact investing, are explored, highlighting their characteristics and advantages in promoting sustainable business practices. The integration of environmental, social, and governance (ESG) criteria into financial decision-making processes and its implications for investment performance and sustainable business outcomes are analyzed. Market dynamics and challenges are assessed, identifying factors driving the adoption of green finance strategies and barriers hindering widespread implementation. Furthermore, the chapter also assesses the environmental, social, and economic impacts of green finance initiatives and illustrates potential effects on sustainable business development. Next, policy implications are discussed, outlining recommendations for policymakers to create an enabling environment for green finance integration and anticipating future trends and developments in the field. Finally, the chapter summarizes the critical role of green finance in advancing the transition to a sustainable economy. It emphasizes the need for collaborative efforts among policymakers, financial institutions, and stakeholders to maximize its potential impact.

Keywords Green finance · Responsible investments · ESG · Sustainable finance · Policy implications

I. Behera (✉)
KIIT School of Management (KSOM), KIIT University, Bhubaneswar, Odisha, India
e-mail: Itishreebehera054@gmail.com

P. Nanda
Parul University, Vadodara, Gujarat, India
e-mail: n.pragyan@gmail.com

H. Baber and M. Fanea-Ivanovici (eds.), *Sustainable Financing—A Contemporary Guide for Green Finance, Crowdfunding and Digital Currencies*, World Sustainability Series, https://doi.org/10.1007/978-3-031-80969-9_6

1 Introduction

The concept of green finance is gaining attention in the rapidly-evolving financial landscape of today. It involves incorporating environmental considerations into economic decision-making, which deviates from traditional financial models. Also, green finance aims to reallocate capital and orient financial strategies towards promoting economic growth and environmental responsibility. This shift in thinking could potentially redefine the purpose of finance, moving away from profit-centric models towards a more sustainable future. The importance of green finance lies in its ability to promote a holistic understanding of success that encompasses economic, social, and environmental dimensions (Al-Sheryani and Haitham 2020; Adhana 2020; Bank 2020). This chapter provides a comprehensive exploration of the transformative potential of green finance. It discusses the theoretical foundations of the connection between green finance and sustainable development, emphasizing that it is not a luxury but a necessity for the continued flourishing of our global society. Green finance challenges businesses to reconsider their roles in the context of a sustainable future and positions itself as a catalyst for systemic change. This chapter invites readers to embrace green finance as a crucial element for sustainable development and work towards a more sustainable future.

Green finance has established itself as a critical driver of sustainable business models, and its significance is apparent in the innovative financial solutions it provides to the economic ecosystem (Bilan et al. 2020). The introduction of green bonds and impact investing has revolutionized the financial industry, providing businesses with powerful instruments to guide them towards a greener, more sustainable future. Green finance has transformed traditional financial theories, proving that economic success and environmental responsibility are interconnected. Green bonds are a type of fixed-income securities designed to fund projects that help the environment. These bonds are effective in directing capital towards sustainable infrastructure initiatives such as renewable energy, energy efficiency upgrades, and sustainable waste management systems (Bhattacharyya 2022). Moreover, traditional finance models refer to the conventional approaches and frameworks used in the financial industry to analyze, evaluate, and make decisions about capital allocation, investments, risk assessment and asset management. These models mostly focus on economic indicators, financial returns and risk management without consideration of environmental or social factors. Green finance promotes sustainable development by considering projects, businesses, and investments with positive environmental or social impacts for capital flows to generate financial returns. It includes investments in sustainable infrastructure, green bonds, energy efficiency, renewable energy, clean technology and socially responsible investments (SRI). Also, green finance incorporates additional criteria related to environmental sustainability, such as water usage, biodiversity impact, carbon emissions and other social factors such as labor standards and community engagement to evaluate the investments. It further seeks to achieve broader environmental and social objectives which include reducing carbon emissions, mitigating climate change, promoting social equity, and enhancing resource efficiency. Specialized financial instruments like green bonds, carbon credits, impact investments and sustainability-linked loans are specifically designed for this purpose. Another essential aspect of traditional and green finance models is risk management. While risk management primarily focuses on financial risks such as market, credit, and liquidity risks in traditional finance, in green finance, it assesses environmental and social risks like climate-related impacts, regulatory changes, reputational risks, and stakeholder concerns to promote long-term sustainable development.

Furthermore, green finance helps in mitigating environmental pollution and its associated health risks by promoting investments in sustainable transportation, clean energy and other pollution reduction initiatives. Thus, green finance contribute to the overall quality of life of communities and improved public health outcomes by reducing toxin exposures and providing cleaner water and air along with improved sanitation and hygiene. Besides, from the social equity perspective, green finance contribute to social stability by supporting affordable housing projects with investments addressing housing shortages, housing disparities and affordability issue in low-income communities. Furthermore, green finance supports financial inclusion by supporting sustainable agriculture or renewable energy initiatives through microfinance programs to empower rural communities and small-scale entrepreneurs promoting economic resilience. Green finance further strengthens social cohesion by involving community engagement and participation while empowering local residents to take ownership in community projects, generate revenue and re-invest the profits in community initiatives and shape their neighbourhood development. Again as discussed previously, green finance provides businesses with diversified investment portfolios that promote positive environmental outcomes. It marks the beginning of a new era of finance where profitability and ecological prudence coexist harmoniously, and businesses become stewards of both economic prosperity and environmental well-being. Some examples of green bonds include the World Bank's Green Bonds and Apple's $1.5 billion green bond offering. The proceeds from the World Bank's Green Bonds are used to finance projects that promote climate change mitigation and adaptation and other environmental objectives. Apple's green bond offering supports the company's efforts to reduce its carbon footprint and transition to renewable energy. Similarly, the Acumen Fund, which invests in social enterprises in developing countries, and the Calvert Foundation, offering investment notes to finance affordable housing and other community development projects showcase some of the examples of impact investing. Impact investing has gained notable attention among investors seeking to generate environmental and social impact alongside the financial returns. Thus, green finance has transformative power and emphasizes the importance of sustainable business models (Canas da Costa and Tobias 2020). It encourages businesses to adopt a long-term approach that prioritizes environmental, social, and economic objectives. Thus, green finance has the potential to transform the financial industry, leading into a new era of responsible investing that benefits society and the environment. The rest of the chapter has been organized as follows: (i) Section 2 examines theoretical foundations of Green Finance; (ii) Section 3 next explores Innovative Financing Solutions; followed by (iii) Section 4 which discusses advantages of innovative green financing solutions; (iv) Section 5 focuses on Integration of ESG Criteria; (v) Section 6 analyzes Impact Assessment and Policy Implications; (vi) Section 7 concludes the discussion.

2 Background

Green finance is a concept that is grounded in the principles of sustainable development. Its purpose is to balance economic growth with environmental protection and social responsibility. Green finance aims to promote sustainable economic growth by investing in environment friendly projects including sustainable agriculture, energy efficient infrastructures and renewable energy (Cui et al. 2020). Moreover, the theoretical foundations of green finance are rooted in the most discussed triple bottom line approach. This approach considers the environmental, economic and social impacts of an investment. It focuses on the balancing of economic growth with environmental and social outcomes, and promoting long-term sustainability. Economic impact includes financial return on investment, job creation, and economic growth. Social impact includes the well-being of the local community, social equity, and stakeholder engagement. Environmental impact includes the reduction of carbon footprint, conservation of natural resources, and protection of ecosystems (Dörry and Schulz 2018, Hafner et al. 2020). Green bonds, green loans, and impact investing are some of the sources of funding for green finance. Green bonds are a type of debt securities issued by companies or governments to fund environment friendly projects (Fagiolo et al. 2020). Additionally, risk management is a key principle of green finance. It assesses and manages the environmental and social risks associated with investments. This includes evaluating the environmental impact of the project, assessing the regulatory and policy landscape, and identifying potential risks and opportunities (Falcone 2020). By doing so, green finance can help minimize environmental and social risks and enhance the financial return on investment.

This approach encompasses a range of financial instruments and strategies designed to promote sustainable practices, mitigate environmental risks, and encourage long-term environmental and social well-being. At its core, green finance extends beyond specific financial products to encompass a broader mindset that values the integration of environmental considerations into investment decisions (Ghosh 2020). Environmental considerations assess the ecological impact of investments, ensuring that projects align with conservation and sustainability goals. Social factors evaluate how investments contribute to community well-being, considering labour practices, human rights, and community engagement. Again, governance criteria focus on the internal structures of companies, promoting transparent and accountable business practices. It emphasizes the importance of responsible financial practices in the face of pressing global challenges such as climate change, biodiversity loss, and social inequality, recognizing the interconnectedness of economic, social, and environmental factors (He et al. 2020).

Traditional financial models, which are often driven by profit, can contribute to environmental degradation and social inequities (Hoshen et al. 2017). In contrast, green finance emphasizes sustainability and recognizes that it is a multidimensional concept that encompasses economic prosperity, social well-being, and environmental stewardship. Aligning financial decisions with these broader objectives is a strategic imperative that positions green finance as a catalyst for positive change (Lee 2020). Green finance's theoretical underpinnings reflect an understanding that addressing environmental challenges and promoting sustainable development require a collaborative effort involving governments, businesses, financial institutions, and society at large. It is a shared responsibility that promotes a collective commitment to responsible financial practices that contribute positively to the overall well-being of the planet and its inhabitants (Liu et al. 2020).

Shifting focus to the financial market dynamics surrounding environmental initiatives, (Sabbaghi and Sabbaghi 2011) provide empirical evidence on the volatility dynamics of carbon emission allowances traded in the United States. Their study on Carbon Financial Instruments (CFI) traded on the Chicago Climate Exchange (CCX) sheds light on the emerging carbon trading market and its implications for environmental policy and financial markets. (Khalil et al. 2022) in their work, highlight the importance of broadening the scope of research beyond carbon emissions and corporate social responsibility (CSR) to gain a deeper understanding of corporate environmental performance. They emphasize the need to include other significant dimensions of environmental sustainability, such as interactions with communities, suppliers, and consumers. Further exploring the relationship between investment and firm performance in another work, (Khalil and Nimmanunta 2022) reveal a U-shaped relationship between knowledge capital and firm performance, indicating that excessive accumulation initially hinders performance until a certain threshold. They also find that innovation spending positively impacts performance but diminish beyond a certain level, highlighting the importance of balanced investment for enhancing firm value. Building upon this perspective, (Khalil and Nimmanunta 2023) further explores into the benefits of companies prioritizing investments in environmentally sustainable initiatives. Lastly, (Arshad and Khalil 2022) contribute to the understanding of the relationship between FDI, carbon emissions, fixed capital accumulation, and economic growth, focusing on Pakistan as an emerging economy. Their findings suggest that while FDI stimulates economic growth, it also leads to increased carbon emissions, posing challenges to environmental quality. The study offers policy recommendations to address these challenges and integrate environmental protection measures into Pakistan's development agenda. The theoretical foundations of green finance can be broadly categorized into three areas:

2.1 Environmental Economics

Environmental economics examines how economic activities impact the environment and natural resources. It aims to promote sustainable development by identifying ways to balance economic growth with environmental protection. Environmental economics further analyzes market failures and externalities that result from economic activities and evaluates the costs and benefits of proposed policy solutions. For example, carbon pricing is a policy measure which reduces greenhouse gas emissions. It further provides businesses and individuals with incentives to encourage carbon footprint reduction through carbon emissions pricing. Similarly, subsidies for renewable energy aim to encourage the adoption of clean energy technologies by making them more affordable and competitive with traditional fossil fuels. Other policy measures that environmental economists recommend include pollution and waste disposal regulations, supports in reducing the negative environmental impact of the economic activities. These policies often involve setting emissions standards or imposing taxes on polluting activities to incentivize businesses to reduce their environmental impact (Louman et al. 2020).

2.2 Socially Responsible Investing (SRI)

SRI is a financial approach considering environmental and social factors when selecting investments. SRI investors evaluate companies based on their alignment with values and its impact on the environment as well as on the society. These criteria mostly include environmental impact, labour practices, human rights, and community relations. Companies engaging in harmful activities like fossil fuel production or weapon manufacturing are typically excluded from socially responsible investment portfolios. Moreover, SRI investors favour companies that are committed to attain both positive environmental and social outcomes, such as support for sustainable agriculture, renewable energy production and gender equality. SRI can take various forms, including positive screening, negative screening and shareholder advocacy. While positive screening actively seeks companies that align with SRI values, negative screening excludes companies not meeting specific criteria (Lüdeke-Freund 2020; Megersa 2020). Thus, in socially Responsible Investing (SRI), investors align their values with investment choices.

2.3 Corporate Social Responsibility (CSR)

Corporate Social Responsibility (CSR) is an important aspect of a company's progress, as it entails the responsibility of companies to act in a way that benefits society and the environment. CSR initiatives involve a comprehensive approach to the company's social and environmental impact, covering practices that minimize adverse environmental effects, promote social justice, and invest in sustainable development initiatives. In this regard, companies may implement various CSR activities, such as reducing carbon emissions, conserving water and energy, supporting local communities, and ensuring ethical supply chains. These initiatives aim to enhance the company's environmental performance, promote social equity, and contribute to society's overall well-being. CSR is now viewed as a strategic imperative for companies, as it can significantly impact their bottom line. Also, companies can attract top talent, boost their reputation, and enhance their financial performance by investing in CSR initiatives (Moro-Visconti 2020).

3 Innovative Financing Solutions

In the domain of sustainable finance as can be seen in Fig. 6.1, innovative financing solutions play a vital role in channelling capital towards environmentally responsible practices. Three notable mechanisms that exemplify this commitment are green bonds, green loans, and impact investing.

Green bonds are a specialized category of debt instruments designed explicitly to fund environmentally beneficial projects. Governments, municipalities, and corporations usually issue these bonds to raise funds for initiatives that promote environmental sustainability (Ng et al. 2021). The use of proceeds spans diverse projects, including developing sustainable energy sources, enhancing energy efficiency, and constructing sustainable infrastructure. The reports include information such as the amount of carbon emissions saved the amount of clean energy generated, and the number of jobs created. Investors' motivations extend beyond financial gains, encompassing Environmental, Social, and Governance (ESG) considerations, aligning with the broader global trend towards sustainable investing. The ESG considerations include corporate governance, climate change and social justice. As the market for green bonds continues to rise, with innovations like social bonds and sustainable bonds, these financial instruments play a crucial role in mobilizing capital for projects that address climate change and promote a more sustainable future. Social

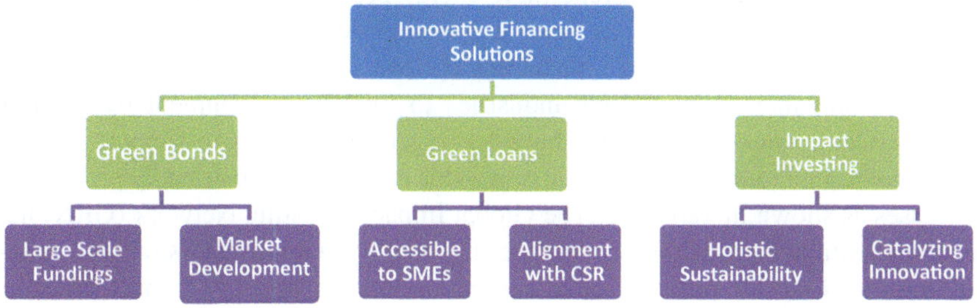

Fig. 6.1 Innovative financing solutions and their advantages

bonds, for example, raise capital to fund social projects such as affordable housing, healthcare, and education. On the other hand, sustainable bonds fund projects that promote environmental and social sustainability (Park and Kim 2020).

Recently, another kind of financial solutions known as green loans have emerged as an innovative financing option for businesses looking to undertake sustainable initiatives. These loans aim to channel funds towards sustainable energy installations, energy-saving upgrades, and other eco-friendly projects. Green loans offer financial support with a clear focus on environmental responsibility, aligning with businesses' commitment to sustainability. The transparency of these loans is an important feature, requiring borrowers to provide detailed information on how the loan proceeds will be utilized and the expected environmental benefits. With businesses prioritizing environmental considerations and stakeholders demanding more sustainable practices, green loans serve as a valuable tool to meet financial and ecological objectives (Sachs 2019). Impact investors are characterized by their active search for opportunities to create measurable impact in areas such as renewable energy, social enterprises, and community development projects (Shipalana 2020). This approach goes beyond a narrow financial lens, taking into account the broader implications of investments on societal and environmental well-being. Impact investing is characterized by its focus on financing projects and enterprises that contribute to sustainable development goals. This includes initiatives that address poverty alleviation, education, healthcare, and access to clean energy. Impact investors prioritize projects that have the potential to create positive change and provide funding and support to organizations committed to making a difference. In the sustainable energy industry, impact investors can fund projects that increase clean energy access, reduce carbon emissions, and contribute to a sustainable energy future (Soundarrajan and Vivek 2016). Social enterprises addressing education, healthcare, or poverty alleviation are attracting impact investment. This funding is provided to organizations committed to positively impacting society. However, as the awareness of environmental, social, and governance (ESG) considerations continues to grow, impact investing is gaining prominence as a strategic approach for individuals, institutions, and businesses.

Green bonds have witnessed significant growth globally, with the Climate Bonds Initiative reporting record issuances in 2021. The market is not confined to specific regions, with active participation from various countries. Notable examples include companies like Apple and governments like Germany issuing green bonds to fund energy-efficient initiatives and renewable energy projects. Similar to green bonds, the market for green loans has been expanding steadily, reflecting businesses' growing interest in sustainable financing options. These loans offer flexibility, enabling companies from diverse industries to finance environmentally friendly projects. Companies like Unilever have leveraged green loans to support sustainability goals, including investments in renewable energy and eco-friendly production processes as shown in Table 6.1. The Global Impact Investing Network (GIIN) has reported continued growth in impact investing activity, with investors seeking financial returns alongside positive social and environmental impacts. Impact investors support projects in various sectors worldwide, including renewable energy, healthcare, education, and social enterprises. For instance, the impact investment firm

Table 6.1 Comparison of innovative financing solutions

Innovative financing solutions	Market size	Geographical presence	Flexibility	Measurement focus	Tangible outcomes	Examples of issuers
Green bonds	Steadily expanding, record issuances	Global	Limited by project type	Environmental impact	Renewable energy, energy efficiency	Governments, corporations (e.g., Germany, Apple)
Green loans	Expanding, increase interest	Global	Flexible, various sustainable projects	Environmental impact	Sustainable production, renewables	Corporations (e.g., Unilever)
Impact investing	Growing with continued increase	Global	Flexible, spans various sectors	Social and environmental impact	Poverty alleviation, social empowerment	Impact investment firms (e.g., Acumen

Acumen has successfully funded projects addressing poverty and social challenges across different regions (Stojanovic and Ilic 2018).

4 Advantages of Innovative Green Financing Solutions

4.1 Advantages of Green Bonds

Green bonds are a type of fixed-income security designed to fund environmentally sustainable projects. By creating a specific avenue for raising funds exclusively allocated to environmentally sustainable projects, green bonds facilitate the implementation of crucial environmental projects that might struggle to secure financing otherwise. Another essential feature of green bonds is their emphasis on transparency and accountability in capital utilization. Issuers of green bonds are often mandated to provide detailed reports on the environmental impact of funded projects (Taghizadeh-Hesary and Yoshino 2020). The alignment with ESG criteria enhances the attractiveness of green bonds as an investment option, driving capital toward environmentally responsible projects. Moreover, green bonds offer investors the benefit of risk diversification by incorporating environmentally sustainable projects into investment portfolios. This strategy allows investors to diversify risk exposure across different sectors and asset classes, mitigating risks associated with traditional investments while contributing to sustainable development. Additionally, including green bonds in portfolios can enhance resilience to environmental risks, strengthening the overall risk-return profile (Volz 2018). Again, adherence to recognized principles and standards, such as the Green Bond Principles (GBP), enhances the credibility and integrity of green bonds. These principles offer a clear framework guiding the issuance, allocation, and reporting of green bond proceeds, ensuring consistency, transparency, and comparability across different issuances.

4.2 *Advantages of Green Loans*

Green loans are becoming increasingly popular among businesses looking for financing options for their sustainability projects. They allow companies to choose from a diverse array of sustainable projects, ranging from energy efficiency upgrades and waste reduction initiatives to the implementation of cutting-edge eco-friendly technologies. This means that businesses can invest in renewable energy infrastructure, adopt sustainable supply chain practices, or enhance environmentally conscious production processes, among other initiatives, and facilitate a customized financing solution that reflects the specific sustainability objectives of the enterprise (Wang et al. 2022). Additionally, opting for green loans also signifies a deeper commitment to sustainability by integrating environmental responsibility into the core operations of a business. Green loans also come with transparency requirements in reporting, where borrowers are obligated to provide detailed information on how the loan proceeds are utilized and the resulting environmental benefits. This transparent reporting contributes to accountability, builds trust between lenders and borrowers, and reinforces the credibility of the business's sustainability efforts. Green loans align seamlessly with the evolving expectations of stakeholders who increasingly prioritize environmentally responsible practices.

4.3 *Advantages of Impact Investing*

Investing with impact is a strategy that aims to generate positive social and environmental outcomes alongside financial returns. One of the key advantages of impact investing is that it dispels the misconception that financial returns must be sacrificed for societal or environmental impact. In addition to its financial impact, impact investing serves as a catalyst for systemic change within the financial ecosystem by advocating for sustainability and responsible practices (Zioło et al. 2020). Another significant advantage of impact investing is its inherent flexibility, which allows investors to allocate capital across a diverse array of sectors. Whether investing in renewable energy, healthcare, education or community development this adaptability enhances the resilience and responsiveness of impact investing to evolving global priorities and emerging opportunities. Therefore, impact investing emerges as a powerful and versatile strategy that demonstrates that financial success and positive societal outcomes can be pursued simultaneously, nurturing a transformative shift towards a more sustainable and responsible global financial landscape.

5 Integration of ESG Criteria

Integrating Environmental, Social, and Governance (ESG) standards into financial evaluation processes represents a transformative shift in the investment landscape, emphasizing the importance of sustainability and ethical considerations. Integration of ESG offers a thorough evaluation of non-financial aspects in investment analysis, providing investors with a complete understanding of risks and opportunities. Prominent frameworks include the (i) Sustainability Accounting Standards Board (SASB), (ii) Global Reporting Initiative (GRI), and (iii) Task Force on Climate-related Financial Disclosures (TCFD). The effectiveness of these frameworks depends on factors such as industry specificity, regional considerations, and the materiality of ESG factors to a particular business. ESG integration has several critical implications for investment performance and sustainable business outcomes. It enables investors to identify companies that demonstrate strong sustainability practices, potentially mitigating risks associated with environmental controversies, social issues, or governance failures. By incorporating ESG factors in investment analysis, companies can identify opportunities and navigate challenges in a rapidly changing global landscape, leading to improved long-term financial performance. ESG integration further emphasizes the importance of three key factors: environmental, social, and governance. The emphasis on environmental factors encourages businesses to adopt sustainable practices, reduce carbon footprints, and embrace renewable energy sources. Social considerations prompt companies to address labour practices, diversity, and community engagement issues, contributing to a positive corporate culture and societal impact. Governance factors ensure transparent and accountable decision-making, reducing the risk of corporate misconduct and enhancing overall corporate governance structures. Thus, adopting ESG criteria into financial decision-making provides investors with a more comprehensive understanding of the risks and opportunities associated with a particular investment. Besides, in recent years, ESG (Environmental, Social, and Governance) criteria integration into financial decision-making processes has gained significant momentum. This development has been driven by increasing interest in responsible and impact investing, with more investors seeking to allocate their capital to businesses that deliver financial returns and demonstrate a commitment to sustainable and ethical practices. This shift in investor preference has led to companies enhancing their ESG performance to attract responsible investors, thereby reinforcing the ESG criteria integration into financial decision-making. However, despite the growing prominence of ESG integration, several challenges must be addressed to ensure effective implementation. One of the primary challenges is the standardization and availability of ESG data, which can vary significantly across industries and regions. This variation is compounded by subjective ESG factors, making it difficult to compare the performance of companies. To address these challenges, stakeholders such as regulators, investors, and companies need to collaborate to establish consistent reporting standards, enhance data quality, and ensure a more robust and standardized approach to ESG integration.

Environmental, Social and Governance (ESG) criteria integration into financial decision-making is needed to promote sustainable and responsible investment practices. However, several challenges and limitations such as data availability, standardization issues and measurement methodologies are definitely associated with it. As data vary widely across all companies and industries, availability of quality data is one of the primary challenges as there is still a lack of reliable and standardized ESG data even though there has been significant progress in recent years. While larger organizations provide comprehensive ESG disclosures, smaller companies fail to do so due to inadequate reporting practices. Again, being qualitative in nature, data related to certain ESG factors like governance practices or social impact are challenging to be collected and quantified for integration. Also, in order to accurately assess the ESG performance of potential investments, investors need standardized, reliable and consistent data. Another major challenge being encountered while integrating ESG into financial decision-making is the lack of standardized ESG metrics and corresponding reporting frameworks across industries and companies. Collaborative efforts from companies, investors, regulators and standard setting bodies are required to address the discussed challenges and revisions. To address the integration challenges and limitations, emerging trends and innovations in ESG integration are evolving continuously which demonstrates the significance of technology-driven solutions in addressing complex challenges in green finance. By incorporating artificial intelligence, block-chain technology and data analytics, companies and investors can enhance accountability and transparency of solutions. Massive amounts of unstructured data are generated from various sources such as social media posts, news articles, and corporate reports due to their heterogeneous nature. Artificial Intelligence and Machine learning algorithms are currently being widely adopted to analyze this vast amount of unstructured data to extract ESG-related insights. Furthermore, to identify relevant ESG trends and events, companies are deploying Natural Language Processing (NLP) techniques for topic modelling and sentiment analysis. Additionally, blockchain technology is being used to enhance traceability and transparency in supply chain of industries with significant social and environmental risks such as agriculture, mining and manufacturing. Again ESG data analytics platforms are emerging which offer comprehensive solutions for collecting, analyzing and reporting ESG data. Data from multiple sources including regulatory filings and public disclosures are aggregated through these platforms to provide the investors with a holistic view of ESG performance. Some of these platforms also allow customization of ESG criteria according to specific industry need so that investors can assess ESG factors more accurately. Thus, these emerging trends and innovations adopt advanced technologies to address complex challenges of sustainable and responsible investing.

6 Impact Assessment and Policy Implications

The adoption of green finance strategies is gaining momentum across the globe due to a growing recognition of the connection between financial activities and environmental sustainability. Investors, corporations, and governments alike have recognized the need to mitigate climate risks and capitalize on emerging opportunities in renewable energy and sustainable technologies. In addition, stakeholders such as customers and employees are becoming increasingly conscious of the environmental impact of businesses, leading to a higher demand for sustainability. The demand for sustainability aligns with the principles of green finance, creating a market-driven push towards more environmentally responsible investment and financing practices. This recognition that sustainable practices contribute to long-term value creation is a key driver propelling the adoption of green finance strategies. Another barrier is the perception of potentially lower financial returns associated with green investments. To overcome this challenge, a clearer demonstration of green investments' financial viability and resilience is needed, dispelling the notion of a trade-off between financial performance and sustainability. The evolving landscape is marked by increasing numbers of financial institutions integrating sustainability into their core operations. Banks, asset managers, and insurance companies are developing green products and services, promoting various financial instruments, including green bonds, green loans, and sustainability-linked instruments. Also, Regulatory frameworks are essential in shaping market dynamics and ensuring accountability through standardized practices. Governments worldwide are introducing and enhancing regulations that encourage or mandate disclosure of ESG-related information, pushing financial institutions to integrate sustainability considerations into their risk management processes. Additionally, regulatory support for green finance, such as tax incentives or subsidies for sustainable projects, further stimulates market growth. Also, stakeholder engagement is a crucial element in the successful implementation of green finance strategies. Investors, businesses, governments, and civil society must collaborate to establish a shared understanding of the importance of sustainability in financial decision-making. Furthermore, the adoption of green finance strategies has significant policy implications. Governments and regulatory bodies play a crucial role in establishing a favourable environment for sustainable finance. One key policy implication is establishing standardized frameworks and metrics for measuring and reporting environmental impact. To encourage sustainability, Governments can promote internationally recognized sustainability standards, such as the Sustainable Development Goals (SDGs) or the Task Force on Climate-related Financial Disclosures (TCFD). In order to assess the sustainability of investment opportunities and promote transparency in financial decision-making, it is crucial to provide investors and other stakeholders with the necessary information.

Some of the recent international initiatives include the Paris Agreement, Task Force on Climate-related Financial Disclosures (TCFD), Sustainable Development Goals (SDGs), European Green Deal and Sustainable Finance Disclosure Regulation (SFDR). The Paris Agreement adopted in 2015 by 195 parties under the United

Nations Framework Convention on Climate Change (UNFCCC) at the UN Climate Change Conference (COP21) in Paris, France is a legally binding international treaty on climate change. The agreement is a great step forward in combating climate change. By aiming to reduce greenhouse gas emissions, it seeks to limit global warming to below 2 °C compared to pre-industrial levels. Additionally, the agreement encourages continued efforts to limit the temperature rise to 1.5 °C, which will help to protect our planet for future generations. The agreement also includes provisions for countries to enhance their climate commitments (National Determined Contributions or NDCs) and promote finance to support mitigation of climate change, strengthen climate resilience and enhance adaptation efforts in developing countries. Thus, the Paris Agreement serves as a core framework for global efforts to combat climate change and promote sustainable development. It emphasizes the significance of mobilizing financial resources from developed countries such as grants, concessional loans and other financial instruments to assist developing countries to implement their climate action plans and building resilience to climate impacts. And green finance closely aligns with the Paris Agreement objectives by directing capital towards investments contributing to climate mitigation, adaptation and resilience. For instance, green finance instruments such as green bonds, climate funds, and carbon finance mechanisms provide funding for renewable energy projects, energy efficiency initiatives, sustainable infrastructure development, and climate resilience measures. These green finance investments help countries transition to low-carbon and climate-resilient economies in line with their nationally determined contributions (NDCs) under the Paris Agreement. Moreover, the Paris Agreement highlights the importance of accountability and transparency in reporting and tracking countries' progress towards their climate commitments.

Another green finance initiative, such as the Task Force on Climate-related Financial Disclosures (TCFD), promotes transparency and disclosure of climate-related financial risks and opportunities. The TCFD, a global organization established by the Financial Stability Board (FSB) in 2015 by then chair of FSB, Mark Carney at the request of the G20 finance ministers and central bank governors, as an initiative to monitor and make recommendations about the global financial system specifically disclosing climate-related financial risks and opportunities. These recommendations provide a framework for companies that enable investors, lenders, insurers and other stakeholders to make better informed decisions. As of in 2023, TCFD has been disbanded by FSB as it is considered that it has fulfilled its role and the International Financial Reporting Standards (IFRS) will be monitoring the disclosures as companies are still needed to disclose TCFD information. The TCFD voluntary recommendations scan through four thematic areas which includes governance, strategy, risk management and finally metrics and targets. In view of these four thematic areas, green finance closely aligns with the TCFD framework recommended for companies. Governance being a key aspect of green finance, companies can demonstrate their commitment to address environmental concerns by describing the board's oversight about climate-related risks and opportunities. In the next thematic area being strategy, categorizing the climate-related risks and opportunities into short, medium and long term helps companies to identify green finance initiative integration areas. Evaluation of the resilience of the organization's strategy under different climate scenarios helps to ensure that green finance investments are resilient to climate-related risks. Environmental risk should be aligned with financial risk to support sustainable allocation of capital in green finance.

Again, the 2030 Agenda providing a global framework for Sustainable Development Goals (SDGs) adopted by all the United Nations' members in 2015 to address around 17 global sustainable development goals including poverty eradication, quality education, reduced inequality, gender equality, clean water, sanitation, clean energy, sustainable cities and communities, biodiversity conservation, peace and prosperity along with the interconnected core environmental, social and economic dimensions. Most of the member countries have aligned to the framework by incorporating the sustainable development goals into national policies and development agendas focusing on promotion of sustainable finance and investments to achieve the goals. The key aspects of the sustainable development goals involve complete approach encompassing social progress, economic prosperity and environmental protection aiming to balance the needs of people, planet and prosperity. Another aspect is the universal application of the SDGs to all countries encouraging them to adapt the goals to its specific context and priorities regardless of their level of development. So other aspects of the SDGs include promotion of social inclusion and equity, integration of environmental sustainability and interconnectedness of various goals and targets. It further integrates ESG considerations into investment decision making promoting responsible and impact-driven investing that supports the SDGs. Green finance also drives innovation and collaboration among financial institutions, governments, businesses and organizations to develop new financial products, tools and initiatives contributing to SDGs. It also encourages organizations to adopt responsible business practices, improve transparency and disclosure, and enhance accountability to stake holders by promoting corporate responsibility and sustainability. Also, policymakers can promote stakeholder engagement to adopt a collaborative approach to sustainable finance. This can include establishing multi-stakeholder platforms and partnerships where investors, businesses, governments, and civil society can come together to establish a shared understanding of sustainability and promote the exchange of best practices. Additionally, policymakers can promote education and awareness-raising initiatives to increase public awareness of the importance of sustainable finance and mobilize support for a more sustainable financial system.

Some potential risks associated with green finance require strong effort from all the stakeholders including the financial institutions, companies, investors, regulators and civil society organizations. Greenwashing is one such risk which refers to the practice of misleading by portraying financial products or activities as environmental friendly when they actually don't meet the required standards. Greenwashing may include misrepresentation of environmental impacts of investments, exaggeration of sustainability credentials, or failure in disclosure of environmental risks involved. Another significant risk encountered by green finance is the misallocation of resources intended to achieve environmental objectives. They may also result in negative consequences. Poorly tailored investments, lack of proper diligence and prioritization of financial returns over environmental impact are some of the reasons responsible for misallocation of resources and failure in effective addressing of environmental challenges. Additionally, these discussed risks can be addressed and mitigated by adhering to strategies that promotes transparency and integrity in green finance. These strategies include clear establishment of standards and certificates along with availability of transparent and credible information to prevent greenwashing. Also, green initiative companies and financial institutions should enhance their disclosure and reporting by providing transparent and comprehensive disclosure of environmental risks, carbon emission data, impacts of methodologies adopted for environmental assessment, energy efficiency based performance metrics. Stakeholders must be engaged in addressing environmental risks to enhance trust and credibility for enhancing accountability and transparency in green finance initiative. Again, Governments and regulators are required to enforce regulations and ensure compliance with disclosure of regulations and standards.

7 Conclusion

The financial industry has recently experienced a significant shift towards green finance, which is rapidly gaining acceptance worldwide. The primary driver behind this trend is the increasing awareness of the effects of climate change on our planet, along with the increasing expectations of stakeholders and the close connection between financial and environmental performance. Green financing solutions have emerged as a powerful mechanism that enables dedicated funding for environmental projects, encourages greater transparency and accountability among businesses, and integrates sustainability into fundamental business operations. Green finance encourages responsible investment practices and facilitates the mobilization of capital towards sustainable development goals. Despite these advancements, further research is imperative to fully understand the potential and impact of green finance. There is a crucial need for the development of precise frameworks that can measure and disclose the environmental impacts of green finance, which is essential for improving transparency and accountability. Overcoming regulatory barriers to broaden access to green financing for a broader range of businesses and investors is paramount. Collaboration among stakeholders is equally vital in maximizing the impact of green finance. Collaborative partnerships help in exchanging knowledge, encouraging innovation, and creating uniform practices and guidelines. Furthermore, examining the influence of green finance on various sectors of the economy, particularly in emerging markets, is critical. The integration of climate-related considerations into investment decisions, driven by climate-related regulations and heightened awareness of climate risks, is expected to continue. Thus, ongoing research should address remaining challenges and shape the future of green finance ethically and resiliently.

References

Al-Sheryani K, Nobanee H (2020) Green finance: a mini-review. Available at SSRN 3538696

Adhana D (2020) Social stock exchange: an innovative financing platform for social enterprises in India. Available at SSRN 3619235

Arshad I, Khalil MA (2022) Environment-growth nexus: revisiting prosperity and challenges associated with foreign direct investment. NIDA Dev J 62(2)

Bank B (2020) Sustainable finance policy for banks and financial institutions. Dhaka: Bangladesh Bank

Bilan YV, Pimonenko TV, Starchenko LV (2020) Sustainable business models for innovation and success: bibliometric analysis

Bhattacharyya R (2022) Green finance for energy transition, climate action and sustainable development: overview of concepts, applications, implementation and challenges. Green Fin 4(1):1–35

Canas da Costa L, Popović T (2020) Financing sustainable infrastructures in a smart cities' context—innovative concepts, solutions and instruments. In: Innovations for metropolitan areas: intelligent solutions for mobility, logistics and infrastructure designed for citizens. Springer, Berlin, Heidelberg, pp 229–243

Cui H, Wang R, Wang H (2020) An evolutionary analysis of green finance sustainability based on multi-agent game. J Clean Prod 269:121799

Dörry S, Schulz C (2018) Green financing, interrupted. Potential directions for sustainable finance in Luxembourg. Local Environ 23(7):717–733

Fagiolo G, Giachini D, Roventini A (2020) Innovation, finance, and economic growth: an agent-based approach. J Econ Interac Coord 15:703–736

Falcone PM (2020) Environmental regulation and green investments: the role of green finance. Int J Green Econ 14(2):159–173

Ghosh K (2020) Role of fintech in green finance and sustainable development. Gyan Manage J 14(2):61–71

Hafner S, Jones A, Anger-Kraavi A, Pohl J (2020) Closing the green finance gap—a systems perspective. Environ Innov Soc Trans 34:26–60

He Z, Liu Z, Wu H, Gu X, Zhao Y, Yue X (2020) Research on the impact of green finance and Fintech in smart city. Complexity 2020:1–10

Hoshen MS, Hasan MN, Hossain S, Al Mamun MA, Mannan A, Al Mamun A (2017) Green financing: an emerging form of sustainable development in Bangladesh. IOSR J Bus Manag 19(12):24–30

Khalil MA, Khalil R, Khalil MK (2022) Environmental, social and governance (ESG)-augmented investments in innovation and firms' value: a fixed-effects panel regression of Asian economies. China Fin Rev Int (ahead-of-print)

Khalil MA, Nimmanunta K (2022) Knowledge capital and innovation across firms in asian economies: a balanced investment approach. Int J Innov Manag 26(10):2250071

Khalil MA, Nimmanunta K (2023) Conventional versus green investments: advancing innovation for better financial and environmental prospects. J Sustain Fin Invest 13(3):1153–1180

Lee JW (2020) Green finance and sustainable development goals: the case of China. J Asian Fin Econ Bus 7(7):577–586

Liu N, Liu C, Xia Y, Ren Y, Liang J (2020) Examining the coordination between green finance and green economy aiming for sustainable development: a case study of China. Sustainability 12(9):3717

Louman B, Meybeck A, Mulder G, Brady M, Fremy L, Savenije H, Trines E (2020) In: Innovative finance for sustainable landscapes, vol 7. CIFOR

Lüdeke-Freund F (2020) Sustainable entrepreneurship, innovation, and business models: integrative framework and propositions for future research. Bus Strateg Environ 29(2):665–681

Megersa K (2020) Improving SMEs' access to finance through capital markets and innovative financing instruments: some evidence from developing countries. Nairobi Securities Exchange. https://www.nse.co.ke

Moro-Visconti R, Cruz Rambaud S, López Pascual J (2020) Sustainability in FinTechs: an explanation through business model scalability and market valuation. Sustainability 12(24):10316

Ng AW, Nathwani J, Fu J, Zhou H (2021) Green financing for global energy sustainability: prospecting transformational adaptation beyond Industry 4.0. Sustain Sci Pract Policy 17(1):377–390

Park H, Kim JD (2020) Transition towards green banking: role of financial regulators and financial institutions. Asian J Sustain Soc Respons 5(1):1–25

Sabbaghi O, Sabbaghi N (2011) Carbon financial instruments, thin trading, and volatility: evidence from the Chicago climate exchange. Q Rev Econ Finance 51(4):399–407

Sachs JD, Woo WT, Yoshino N, Taghizadeh-Hesary F (2019) Importance of green finance for achieving sustainable development goals and energy security. Handbook Green Fin 3

Shipalana P (2020) Green finance mechanisms in developing countries: emerging practice. Covid-19 Macroecon Policy Responses Africa 2:1–19

Soundarrajan P, Vivek N (2016) Green finance for sustainable green economic growth in India. Agric Econ/Zemědělská Ekonomika 62(1)

Stojanovic D, Ilic B (2018) Green financing in the function of risk management environment and sustainable economic growth. In: Economic and social development: book of proceedings, 69–76

Taghizadeh-Hesary F, Yoshino N (2020) Sustainable solutions for green financing and investment in renewable energy projects. Energies 13(4):788

Volz U (2018) Fostering green finance for sustainable development in Asia. In: Routledge handbook of banking and finance in Asia. Routledge, pp 488–504

Wang KH, Zhao YX, Jiang CF, Li ZZ (2022) Does green finance inspire sustainable development? Evidence from a global perspective. Econ Anal Policy 75:412–426

Zioło M, Bąk I, Cheba K, Spoz A (2020) The relationship between banks and company business models-sustainability context. Procedia Comput Sci 176:1507–1516

Challenges Faced by Social Entrepreneurs in Funding Their Innovative Solutions

Daria Elisa Vuc

Abstract Modern times come with greater challenges for the society and economy altogether. Anyone nowadays can easily name social issues that are progressively drawing our attention: whether we think of environmental aspects, social inequality, immigration, health care, homelessness, unemployment, racism and so many more, it is where social entrepreneurs step in. As agents of change in the society, entrepreneurs can boldly tackle the world's most common problems through creating business models which break new ground for social responsibility. However, although we are witnessing a rise in the social enterprises globally, one of the principal obstacles these initiatives face remains access to funding. The objective of this chapter is to explore how social entrepreneurship is encouraged to grow and stay sustainable with the means of sustainable financing. While crowdfunding continues to be the most well-known source of funding for the organizations in question, some other options are available. More focus should be attached to the topic of financing in social entrepreneurship due to the fact that this sector creates significant opportunities both for the issues it addresses to and for the growth of the economy in general. As social entrepreneurs need investments, this chapter unveils the most sustainable financing methods for them.

Keywords Social entrepreneurship · Sustainable financing · Crowdfunding · Social investment · Social value · Sustainability

1 Introduction

Throughout the history of all times, humans had to face many challenges which shaped the world as we know it now and still have to face modern issues. Most of the countries in the world deal with problems such as poverty, wars, pollution, child

---

D. E. Vuc (✉)
Doctoral School of Economics I, Bucharest University of Economic Studies, 6, Piata Romana, District 1, 010374 Bucharest, Romania
e-mail: daria.vuc@economie.ase.ro

H. Baber and M. Fanea-Ivanovici (eds.), *Sustainable Financing—A Contemporary Guide for Green Finance, Crowdfunding and Digital Currencies*, World Sustainability Series, https://doi.org/10.1007/978-3-031-80969-9_7

labor or unemployment. Innovative solutions to all these challenges can be found and fostered by institutions, NGOs, social organizations and social entrepreneurs.

Entrepreneurship stimulates the well-being of the society through creative ideas which generate benefits for consumers and sustains the economic growth of countries. Besides that, entrepreneurship has an incredible power to change the society. This is the reason why social entrepreneurship was developed: to tackle more complex social issues by generating value for the affected parts of the society. Many definitions were attributed to this special type of entrepreneurship, but what all of them want to embody is the capacity of social entrepreneurs to objectively identify a part of the society that is not working well and the desire to provide solutions to make it better (Harding 2007). What sets social entrepreneurship apart from traditional entrepreneurship is the end goal mainly, which is to create value and not necessarily to generate revenues. However, although social enterprises do not focus on the monetary outcomes, they still need financial investments to develop, grow and stay in business. There are some potential sustainable financing alternatives to gain capital for social businesses.

Thus, this chapter aims to briefly analyze what some of these sustainable financing methods are available for social enterprises and how each of them can impact the success of such an organization to complete its goal of solving the segments of society that need adjustments over a longer period of time.

2 Background

Social entrepreneurship can be powerful when it comes to creating solutions to issues that our modern society is dealing with. The 17 Sustainable Development Goals offer business opportunities for social entrepreneurs who want to make a positive change in the world by tackling problems like poverty, environmental pollution, gender inequality, water scarcity, unemployment, homelessness, poor education, and the list may go on. While incredibly innovative and creative ideas may emerge out of social entrepreneurs' sharp minds, any social organization needs financial means to start from scratch and to operate in the first place, even though money is not their end goal. At the moment, with the rapid technological advancements, there are some available sustainable financing alternatives for social organizations, one of the most popular being crowdfunding. However, this chapter explores more methods for social entrepreneurs to raise money for their social projects, and parallels between traditional financing methods and more innovative and sustainable ones are drawn for the benefit of the business heroes who can make an impact on the society.

3 Social Entrepreneurship and Socially Responsible Investment (SRI)

Poverty, social inequality, homelessness, discrimination, violence, unemployment, exploitation, gentrification, digital divide or ageism are just some of the issues that negatively affect the current society. Social entrepreneurship is an emerging topic and it is strongly related to sustainability because its orientation is a long-term one and its goals should be relevant according to the 17 UN Sustainable Development Goals (SDGs). Social businesses can fight for no poverty, zero hunger, good health and well-being, quality education, gender equality, sanitation, affordable and clean energy, decent work and economic growth, industry, innovation and infrastructure, reduced inequalities, sustainable cities and communities, responsible consumption and production, climate action and peace, justice and strong institutions (UN 2015a, b).

As agents of change, social entrepreneurs, in contrast to traditional business people, are motivated not necessarily to make money but are driven by an inner passion to change the world. Despite this, financing still represents an extremely important factor in the development of social organizations. The practice of investing money in projects and businesses that bring a positive social impact in the world is called 'socially responsible investing' (Investopedia 2022). This type of investing can be made directly to a social enterprise that was unable to gain money from traditional sources (like banks or financial institutions) or by the means of mutual funds or exchange-traded funds (ETFs). As the capital market is being diversified, socially responsible investing becomes essential both for investors who wish their money to be involved in socially-oriented projects and for entrepreneurs, likewise (Michalczuk and Konarzewska 2020). In the past, social investment was called ethical investment, and was considered to be intended for a niche market which included organizations that fought for activism in the society (Revelli 2016). Investors need to pay attention to what causes they finance, and they have to choose to invest in organizations that conform with the environmental, social, and governance (ESG) criteria (Scholtens 2014).

Creative thinkers need the financial means to put out their innovative ideas based on market-approach techniques to reach the targeted segment of stakeholders for whom their social initiatives create social value. Research in the field (Parekh and Attuel-Mendes 2022) has stated that social businesses can access more options for gaining funds from multiple and various sources, rather than limit to traditional options such as banks or impact investors.

In the following sections, the funding methods of a social enterprise are briefly analyzed and explained, so that some of the existing sustainable financing methods are better understood.

4 The Importance of Sustainable Financing Methods in Social Entrepreneurship to Create Social Value Long-Term

While feasible solutions can be created in unimaginable creative and innovative ways and even though money is not the principal aim of the social entrepreneurial activity, it should be generally agreed upon the fact that entrepreneurs still need funds to put their ideas into practice. Governments have strategies to encourage individuals to set up social enterprises and, besides that, there are also several suitable financing methods for enabling people with vision to implement their ideas (Harding 2007). Market-based approaches to create value for stakeholders are used in social entrepreneurship as well as in any kind of entrepreneurship and money remains an important means to achieve social value and innovation. Nowadays, with the help of technological advancements and breakthrough innovations in finance, it is quite easier for social enterprises to raise money in order to develop, grow, stay in business and become sustainable. Social organizations can obviously opt for traditional financial institutions/banks, but as research has grown in this area, other more feasible alternatives are continually discovered, despite the old ones which are not as sustainable. It has been argued through impact investment education programs whether money can generate financial returns and social value at the same time (Parekh and Attuel-Mendes 2022).

Nonetheless, one thing is certain: money should bring a social impact. In this chapter, some of the sustainable available financing options will be presented with regards to encouraging social entrepreneurs to develop sustainable business models for their ideas.

The reason why it is more difficult for social enterprises to obtain investments is the underlying risk of generating profit and paying back the investors, due to the fact that social value metrics may not be properly put in place and, as a result, the deep-rooted mission of the organization cannot be well-perceived by stakeholders in its entirety (Biancone and Radwan 2018). Moreover, social enterprises create jobs and add to the growth of the economy, besides fighting against social issues, but it is hard for them to survive if financiers are not given the proper information in order to become interested and convinced to invest (Dudley 2021).

4.1 Is Traditional Banking Sustainable for Financing a Social Enterprise?

While a normal loan can be accessed by any entrepreneur who wants to build and grow their business, traditional banking is nowadays challenged by more sustainable and innovative financial methods when it comes to helping social organizations develop their business models into successful enterprises. Due to the fact that banks are being heavily regulated and have particular legal structures, the process of obtaining a traditional loan may become extremely lengthy and rigorous and might even end up in not being granted (Kumar et al. 2024).

Start-ups in general present a very high risk for conventional banks in terms of not being able to achieve the financial goals, because they are in the early stages when the certainty of returns on the investments is relatively low. Despite how innovative some social projects may be, around 20% of enterprises in the European area abandon them because of the lack of opportunity to gain bank finance (Coronel-Pangol and Paule-Vianez 2024).

Hence, typical sources of investment are prone to be shadowed by substitutes that openly comprehend the prevailing dynamics of social crises, and traditional banking has been slowly replaced by crowdfunding platforms that encourage social entrepreneurs to contribute to the creation of jobs and growth of the economy through their innovative solutions (Kumar et al. 2024).

4.2 Types of Investors and Their Impact on Social Enterprises' Missions

Although enterprises devoted to social missions mostly used to look to charity as a traditional method of financing, they can access more capital now with the help of other innovations in the financial sector. Social enterprises can be both for-profit and nonprofit ventures. Each type needs to cover its costs in order to grow and expand its social benefits into the world. Some of the social enterprises manage to earn a sufficient profit through the selling of goods and through providing services at a premium price such as for organic food or green energy (Bugg-Levine et al. 2012), while other ventures survive only through government subsidies, charity and donations from individuals. Nonetheless, the lack of necessary money to fund social missions results in a lot of time and effort consumed from the social entrepreneurs' side—time and effort that could be focused on the core missions of their businesses. Once the potential of social enterprises generating financial returns is drawn to the investors' attention, more money will be available to such organizations (Bugg-Levine et al. 2012). For instance, donors do not expect to get a return from their investment because the organizations are then considered to be successful if the social mission is reached accordingly.

However, more financial opportunities can be reached through innovative financing methods and three examples from this cited article (Bugg-Levine et al. 2012) are: loan guarantees, quasi-equity debt for nonprofits which are not able to benefit from equity capital, pooling funds or social impact bonds that enable governments to offer funds for public projects. Obviously, venture capitalists (VCs), angel investors and impact funds remain sources of financing as well, especially in the first stages of development of the social organization, when it primarily needs scaling to grow in the market. Thus, equity and investment not only provide enterprises with money, but they add value through offering access to markets, mentorship and strategic partnerships for visibility and growth altogether.

4.3 Microlending in Social Enterprises

Microlending is another financial service built as an arrangement between the lender and the borrower which refers to savings, small loans, insurances or payment services. A potential downside of this service lies in the transaction costs it creates for both parties: on the one hand, the lender has to look for loanable funds, then design credit arrangements and finally evaluate the feasibility of the project and if loan applications should be approved; on the other hand, the borrower must spend significant time with the loan application and even undergo collateral evaluations (Simba et al. 2023). Due to the fact that the main activity and mission of social enterprises is to discover innovative solutions to social issues, this type of financing may not be exactly suitable for scaling a social organization that may not be able to fulfill the formal requirements of the lending institution. Shortly, microlending on its own might not be perceived as a sustainable type of finance for social organizations.

The term 'microlending' is also used in crowdfunding, an alternative financing method which will be discussed later in the next subchapters, and it refers to the person-to-person lending method of raising enough money to scale a social enterprise (Braund and Schwittay 2016).

4.4 Stages of Funding, Grants and Impact Investing

The word "equity" has been mentioned a few times until now and it is not breakthrough information that it is essential for a company to stay in business and for the financiers to be interested in it. In the pre-seed stage, when the organization is at the beginning of its activity, grants, using personal savings or resources (bootstrapping) and the money from family and friends can give a good start for the social entrepreneur. Funding through grants[1] can be explained by going for philanthropic and charity foundation programs or even government funds for local projects. Nonetheless, the pre-seed stage funding can be tricky because involving family's money in the business does not always necessarily result in an expected outcome and, where family subjectivity intervenes, many more misunderstandings and improper alignment of ideas and visions may alter the core mission of the business.

In the next stage, which is the seed stage, the social business already has a model and a value proposition and the risks involved have been already assessed. In this case, angel investors and crowdfunding are the most common options. And, as impact

[1] https://socialbusinessdesign.org/what-funding-sources-for-social-businesses-and-social-enterprises/.

investing can be correlated with crowdfunding, let us see its implications in the social enterprises. Over the years, research in impact investing has been growing and it is continually observing the social trends that require money support. Impact investing[2] takes place when someone makes an investment with the hope of contributing to social and environmental issues and, at the same time, gaining financial returns on those investments. It is collectively believed that it is all about making an impact. Whether we think of impact investing as a tool to donate to social projects and nonprofits, crowdfunding helps it increase in scale. Investors, for instance, need to be subjectively connected to the cause they believe in: if you are an investor and want to have a greener lifestyle and decide to reduce the usage of fossil fuels, you might wish to find organizations to invest in that focus on developing cutting-edge renewable energy solutions.

5 Crowdfunding in Social Enterprises

While crowdfunding platforms are really well-suited to raise awareness among high-net-worth individuals who can finance social solutions through impact investment,[3] social media platforms play a vital role in publicizing those organizations so that they can reach the specific investors. Since the example of renewable energy was given before, one study (da Silva et al. 2023) explored how green technologies have broader access to funding than other organizations and that social impact investors have more information about environmental problems and place less importance on the financial returns.

5.1 General Aspects of Crowdfunding and Ways in Which It Helps Social Enterprises Grow

Crowdfunding, or crowdsourcing, can be of three types: (1) reward-based; (2) debt-based (known as peer-to-peer lending) and (3) equity-based, all relating to public cooperation and involvement. At this stage, we are talking about the third stage that a social business can reach by growing and scaling and by having a well-stated value proposition, a very strong and validated business model and a proven track record. No matter the type, a crowdfunding platform plays the role of a collaborative Internet-based intermediary between the funder (or the investor, the giver) and the creator of the project (or the social entrepreneur), thus a harmonious social network

---

[2] https://www.fidelitycharitable.org/guidance/philanthropy/impact-investing.html#:~:text=Impact%20investing%20is%20the%20act,benefits%20while%20generating%20financial%20returns.

[3] Good practice: Social enterprise—social finance—crowdfunding will help impact investing to grow (2015). *Third Sector*, 62. Retrieved from https://www.proquest.com/trade-journals/good-practice-social-enterprise-finance/docview/1693329683/se-2.

being created around an altruistic devotion to stand up for a noble idea. Basically, crowd refers to a dispersed external large audience which consists of numerous individuals who each offer a small amount of money, a practice that can fast add up to a significant social investment (Belleflamme et al. 2014). In the context of the actual digital era, social crowdfunding is associated with an open call for gaining funds easily with the help of the internet, by instantly collecting necessary data on donations and contributions for the social purposes.

Quite interestingly, research (Bernardino and Santos 2016) has been conducted on how the personality traits of social entrepreneurs might determine which type of funding they choose for their organizations. So, according to the mix of internal and external drivers that encourage such entrepreneurs to decide to solve the challenging problems of our society, intuition becomes an important variable in the willingness to experience and draw very important values such as the vision to fulfil the basic human needs, the need for integrating sustainability for a healthy long-term development of the society based on the interconnectivity between individuals and businesses, social networks, creativity and innovation and financial results.

Thus, the social crowdfunding project aims not only to raise capital for a social organization, but to also raise awareness through societal involvement and interaction. While traditional crowdfunding projects follow the lending model, which means that "the giver" (i.e. the crowdfundee) expects a return on the invested amount of money, social crowdfunding bases its model on the "patronage" model (Bernardino and Santos 2016), in which funders become philanthropists who do not expect a monetary return on their contributions; they rather want to see the social mission become a reality because they empathize with an almost similar feeling of making an impact in the world. Moreover, social organizations are able to benefit from higher donations because donors are prone to offering recurring funds especially when they receive updates on the progress of the social project.

5.2 *Successful Crowdfunding and the Community Built Around It Through Platforms*

For an investor, it all comes down to the feeling that the offered money contributes to solving a social problem that the world is facing. What is more attractive about social crowdfunding is the community that is created around it. By means of the digital platform, a social project attracts donors with identical or complementary interests, and as a consequence, donations increase. Non-profit oriented and social projects are more likely and more successful (Moritz and Block 2014) in receiving crowd funds because their mission's credibility outstands other businesses' and, again, the personal qualities of a social entrepreneur and emotional triggers can influence how the social business is seen out there in the world. The funding access and success are influenced by the creativity that is used to shape a solution for a legitimate social problem and social entrepreneurs, as opposed to commercial ones, need to communicate their mission and vision in such a way that money flows naturally into the social project through the crowdfunding platform (Calic and Mosakowski 2016).

However, the funding success may be hindered by social entrepreneurs' lack of proper business experience and related education in the field, because they tend to focus more on the non-financial outcome of their ideas rather than to explore the potential of financial liquidity, an aspect that investors are mostly interested in. This is the reason why crowdfunding is so effective for non-profit projects that fight for a cause—the community built around it composed of people who believe in the growth and success of a social organization or project sustains initiatives which have the power to send a message to more persons who would have the chance to discover such projects and invest in them in their turn. Subsequently, the most important legislation on equity-based crowdfunding exists in the United States (Hörisch 2015), as policy-making institutions have started to understand the importance of offering support to social entrepreneurs who look for financial aid. Furthermore, some projects, especially the non-profit ones, may represent a high risk for a traditional bank to offer a loan, because of an idea's certainty to stay in business and prosper, and this is where crowdfunding intervenes and saves social entrepreneurs from the hardship of finding investors and getting them to believe in them (Rey-Marti et al. 2019).

Risks exist in the case of crowdfunding platforms as well and they can be represented by cybersecurity aspects or they can revolve around a project's good faith. A project's good faith resides in the fact that it is not a fraudulent scheme and that its scope sticks to a normative legitimacy translated into an organization's consistent and trustworthy behavior with the values it creates for a normal society (Rey-Marti et al. 2019).

Another positive aspect of crowdfunding is the lack of the common bureaucracy met with financial institutions when applying for a credit and the transparency it provides to gain capital (Bilgin et al. 2018). There is a tendency among scholars (Lehner and Nicholls 2014) to describe crowdfunding as a democratic method of financing as it is the one alternative that tailors the needs of a social entrepreneur to be bold enough to go out in the world and commit to saving a small part of it and the interest of people who believe in non-governmental entities' potential in delivering social services.

6 Conclusion

At only a click distance, anyone could navigate the Internet and find out that the United States rank the first as the country where crowdfunding is accessed as an alternative means of financing for social investments, followed by European countries whose implementation of the crowdfunding concept is on the rise. Although social entrepreneurship is somewhat a young and still developing research field, there are many opportunities for passionate entrepreneurs to tackle on grandiose missions and be able to benefit from financial investments to support their business ideas, as long as alternative financing will be enhanced and encouraged among peers.

Social entrepreneurs have to deal with numerous challenges in their road to success, from coming up with feasible ideas to accessing enough funds to sustain them. Traditional banking and microlending may not be effective methods to gain those funds because of the risks involved and the long process of granting loans. Impact investing, angel investors, venture capitalists, social grants, Initial Public Offering (IPOs) are some of the possible financing alternatives for social organizations. However, crowdfunding not only remains the most well-known and most user-friendly source of money for such investments, but it will continue to contribute to the economic growth of regions and countries because it helps social entrepreneurs to sustain solutions for the parts of the world or society that are damaged.

Terms like "charity" and "philanthropy" will be soon replaced by smart investing in social businesses, but only time will tell. Social entrepreneurship financing remains at the intersection of moral obligation and empathy (Ukil et al. 2024) found in both the creators of solutions for global and local issues and the investors who have the financial means to help sustain a prosperous world. Institutions which support social entrepreneurial initiatives need to understand the values that are behind the ideas and solutions, and ethical banking, social incubators and hubs and crowdfunding are legitimate funding solutions in this context (Simón-Moya and Rodríguez-García 2021).

More importance should be invested in finding sustainable financing solutions for social enterprises and social innovations, because, despite the revenue not being the main objective of their activity, such organizations are able to create jobs, foster the economy and touch sensitive issues that most usual profit-making companies do not encounter in their activities, no matter how many Corporate Social Responsibility (CSR) initiatives they may embark on. However, corporations can and are encouraged as well to offer non-repayable funds to support social projects or similar initiatives that may align with the organizational culture's goals and values.

As mentioned in the beginning, not all heroes are entrepreneurs, but all social entrepreneurs must be a special kind of heroes. With the rapid, ingenious technological advancements and an increased focus from the academia, breakthrough innovations will be witnessed in the financing field over the time in most of the economic activities, especially in the social entrepreneurship sector.

Competing Interests The author has no conflict of interest to declare that is relevant to the content of this chapter.

References

Belleflamme P, Lambert T, Schwienbacher A (2014) Crowdfunding: tapping the right crowd. J Bus Ventur 29(5):585–609. https://doi.org/10.1016/j.jbusvent.2013.07.003

Bernardino S, Santos JF (2016) Financing social ventures by crowdfunding: the influence of entrepreneurs' personality traits. Int J Entrepreneursh Innov 17(3):173–183. https://doi.org/10.1177/1465750316655903

Biancone PP, Radwan M (2018) Social finance and unconventional financing alternatives: an overview. EJIF Euro J Islamic Finance 10. https://doi.org/10.13135/2421-2172/2818

Bilgin M, Danis H, Demir E, Can U (2018) Crowdfunding as a model and financing instrument in social enterprises. Consum Behav Org Strat Fin Econ Eurasian Stud Bus Econ 9:203–220. https://doi.org/10.1007/978-3-319-76288-3_15

Braund P, Schwittay A (2016) Scaling inclusive digital innovation successfully: the case of crowdfunding social enterprises. Innov Dev 6(1):15–29. https://doi.org/10.1080/2157930X.2015.1067392

Bugg-Levine A, Kogut B, Kulatilaka N (2012) A new approach to funding social enterprises. Harvard Bus Rev. Retrieved from https://hbr.org/2012/01/a-new-approach-to-funding-social-enterprises

Calic G, Mosakowski E (2016) Kicking off social entrepreneurship: how a sustainability orientation influences crowdfunding success. J Manage Stud 53(5):738–767. https://doi.org/10.1111/joms.12201

Coronel-Pangol K, Paule-Vianez JO-C (2024) Conventional or alternative financing to promote entrepreneurship? An analysis of female and male entrepreneurship in developed and developing countries. Int Entrepreneursh Manage J 20:163–187. https://doi.org/10.1007/s11365-023-00906-w

da Silva Fernandes E, Grochau IH, Schwengber Ten Caten C (2023) Impact investing: determinants of external financing of social enterprises in Brazil. Sustainability 15(15):11935. https://doi.org/10.3390/su151511935

Dudley E (2021) Social capital and entrepreneurial financing choice. J Corp Fin 70:102068. https://doi.org/10.1016/j.jcorpfin.2021.102068

Harding R (2007) Understanding social entrepreneurship. Industr Higher Educ 21(1):4–107. https://doi.org/10.5367/000000007780222723

Hörisch J (2015) Crowdfunding for environmental ventures: an empirical analysis of the influence of environmental orientation on the success of crowdfunding initiatives. J Clean Prod 107:636–645. https://doi.org/10.1016/j.jclepro.2015.05.046

Investopedia (2022) Guide to socially responsible investment (SRI). Retrieved from https://www.investopedia.com/terms/s/sri.asp

Kumar J, Rani M, Rani G, Rani V (2024) Crowdfunding adoption in emerging economies: insights for entrepreneurs and policymakers. J Small Bus Enterpr Dev 31(3). https://doi.org/10.1108/JSBED-05-2023-0204

Lehner OM, Nicholls A (2014) Social finance and crowdfunding for social enterprises: a public–private case study providing legitimacy and leverage. Int J Entrepr Fin Fin Risk Account Perspect 16(3):271–286. https://doi.org/10.1080/13691066.2014.925305

Michalczuk G, Konarzewska U (2020) Socially responsible investing in the European Union's capital markets. Ekonomia i Prawo Econ Law 19(4):756–771. https://doi.org/10.12775/EiP.2020.049

Moritz A, Block JH (2014) Crowdfunding: a literature review and research directions. SSRN Electron J. https://doi.org/10.2139/ssrn.2554444

Parekh N, Attuel-Mendes L (2022) Social entrepreneurship finance: the gaps in an innovative discipline. Int J Entrepreneurial Behav Res 28(1):83–108. https://doi.org/10.1108/IJEBR-05-2021-0397

Revelli C (2016) Re-embedding financial stakes within ethical and social values in socially responsible investing (SRI). Res Int Bus Fin 38:1–5. https://doi.org/10.1016/j.ribaf.2016.03.003

Rey-Marti A, Mohedano-Suanes A, Simón-Moya A (2019) Crowdfunding and social entrepreneurship: spotlight on intermediaries. Sustainability 11(4). https://doi.org/10.3390/su11041175

Scholtens B (2014) Indicators of responsible investing. Ecol Ind 36:382–385. https://doi.org/10.1016/j.ecolind.2013.08.012

Simba A, Ogundana OM, Braune E, Dana L-P (2023) Community financing in entrepreneurship: a focus on women entrepreneurs in the developing world. J Bus Res 163. https://doi.org/10.1016/j.jbusres.2023.113962

Simón-Moya V, Rodríguez-García M (2021) Sources of financing for social entrepreneurship. In: The emergence of social entrepreneurship. Contributions to management science. Springer, Cham, 29–39. https://doi.org/10.1007/978-3-030-80635-4_4

Ukil MI, Almashayekhi A, Ullah MS (2024) Feeling compassion and responsible but not starting a social venture: role of empathy and moral obligation in social entrepreneurial intention. Soc Bus Rev 19(1):132–154. https://doi.org/10.1108/SBR-09-2022-0227

United Nations (2015a) United Nations. Retrieved from https://sdgs.un.org/goals

United Nations (2015b) Retrieved from Department of Economic and Social Affairs. Sustainable Development. https://sdgs.un.org/goals

Crowdfunding—A Transformative Force in Sustainable Finance

Archana Yadav, Faizan Ashraf Mir, Arif Hasan, Vandana Kushwaha, and Shubhangi Vitthal Gaikwad

Abstract Crowdfunding is a sustainable funding option that involves raising small amounts of money from many people, typically via online platforms or social networks. Crowdfunding offers a decentralized approach to raising funds for entrepreneurs by connecting with a wider audience of potential backers who intend to contribute towards a specific idea. This chapter delves into the world of crowdfunding by defining the concept of crowdfunding and its historical evolution since the pre-internet era. The chapter discusses various types and models of crowdfunding and their subsequent efficacies. It discusses the transformative role of crowdfunding platforms and their impact on the entrepreneurial culture. The discussion encompasses the role of crowdfunding on community engagement and social impact. The chapter concludes by presenting various inherent challenges and risks involved in crowdfunding, thereby providing a comprehensive overview of the dynamics that shape this innovative funding landscape. The chapter provides an in-depth understanding into the evolution, functioning and potential ramifications of crowdfunding in the contemporary financial ecosystem.

Keywords Crowdfunding · Alternate funding · Entrepreneurship · Community engagement

---

A. Yadav
GLA University Mathura, Mathura, UP, India

F. A. Mir (✉)
Central University of Kashmir, Tulmulla, J&K, India
e-mail: faizan.ashraf@cukashmir.ac.in

A. Hasan
ABS, Amity University, GwaliorMadhya Pradesh, India

V. Kushwaha · S. V. Gaikwad
School of Economics and Commerce, MIT World Peace University, Pune, India
e-mail: shubhangi.gaikwad@mitwpu.edu.in

H. Baber and M. Fanea-Ivanovici (eds.), *Sustainable Financing—A Contemporary Guide for Green Finance, Crowdfunding and Digital Currencies*, World Sustainability Series, https://doi.org/10.1007/978-3-031-80969-9_8

1 Introduction

In the bustling marketplace of ideas, where imagination meets possibility and dreams take flight with wings, there exists a stage unlike any other: **The Crowdfunding Platform** (Wachira and Wachira 2021). This is not your typical investor boardroom or pitch deck competition. Here, the audience is the investor, the funding comes in dribs and drabs from a global community. The currency of exchange is passion, creativity, and a whole lot of storytelling. Crowdfunding is not just about securing funds; it is about building a community (Ryu and Suh 2021). Your backers are not just anonymous faces on a screen; they are your early adopters, your brand ambassadors, and your biggest cheerleaders. Engaging with them, keeping them updated, and delivering on your promises are crucial to fostering loyalty and ensuring that your campaign thrives. Crowdfunding has become the democratization of dreams a platform where anyone with a compelling vision can bypass the gatekeepers and directly connect with the very people who believe in their cause (Galuszka and Brzozowska 2017). It is a melting pot of ideas, from tech startups to artistic endeavours, social impact projects to quirky inventions, all vying for the support of a crowd that's hungry for innovation and eager to be part of something bigger. But navigating this vibrant chaos requires more than just a catchy pitch and a polished online profile.

The financial crisis of 2007–2008 saw the advent of crowdfunding as a disruptive financing tool, which later on got a boost during the COVID-19 pandemic (Baber and Fanea-Ivanovici 2022). The academic exploration of this novel financing instrument, i.e., crowdfunding, began in the year 2010 and has been continuing since then. This untapped scientific territory has drawn significant interest among researchers and academicians across the globe with the U.S. (25.7%) being the top contributor in terms of academic papers during the period from 2006–2021 followed by China (14.3%) (Baber and Fanea-Ivanovici 2022). The concept, nature, evolution and its effectiveness have been studied over the years more so from the perspectives of Business, Management and Accounting (45%) followed by Computer Science (29.2%) (Baber and Fanea-Ivanovici 2022).

In today's age, merely showcasing a product on a prominent platform is an effective marketing strategy. Crowdfunding provides a platform for entrepreneurs to present their product in the public eye, and even if the product or idea fails to get investment from the backers, it still provides valuable insights to the entrepreneurs to further polish the idea or drop the idea, thereby acting as marketing testing labs for young entrepreneurs. (Fanea-Ivanovici and Baber 2021a, b). Crowdfunding has evolved as a viable substitute, especially for creative entrepreneurs who otherwise would find it difficult to raise capital through traditional financing channels (Baber and Fanea-Ivanovici 2023). It has become a very convenient source of finance for not only well-established businesses but also inexperienced entrepreneurs and start-ups (Baber and Fanea-Ivanovici 2021a).

Sustainability is a cross-cutting concept with a wide range of implications, especially for social and environmental development. An increasing number of publications on crowdfunding and sustainability have found a clear relationship between the two variables (Martinez-Climent et al. 2019).

You need a strategy, a deep understanding of the various crowdfunding models and the ability to tap into the hearts and minds of your potential backers to stand out in the crowd truly. Crowdfunding has emerged as a transformative force and reshaping how businesses and creative projects access capital in the ever-evolving landscape of entrepreneurial finance (Cumming and Groh 2018). This chapter explores the dynamics, mechanisms and implications of crowdfunding as a sustainable alternative funding option. From its historical roots to the contemporary platforms that define its current state, we delve into the multifaceted world of crowdfunding and its impact on innovation, entrepreneurship and community engagement.

2 Background

To understand the contemporary significance of crowdfunding, we embark on a journey through its historical evolution. Tracing its roots back to informal community-based fundraising, we explore how crowdfunding has evolved over the years, driven by technological advancements and changing socio-economic landscapes (Zhang and Chen 2019; Gupta et al. 2023a, b). From early examples of artistic patronage to the emergence of online platforms, the chapter provides a comprehensive overview of crowdfunding's journey. It is a financing model which has undergone a fascinating historical evolution, transforming from localized community initiatives to expansive global platforms. This journey spans centuries, reflecting changes in technology, communication and the democratization of finance.

Crowdfunding's roots can be traced back to the pre-internet time (pre-2000s), which is community-based initiatives and grassroots fundraising efforts (Lee and Chiravuri 2019) Communities would come together to support local projects, causes, or individuals in need. Whether it was pooling resources for a community project or supporting an artist's work, these early forms of crowdfunding were often localized and relied on personal connections within small communities (Daniele 2021). The advent of the internet marked a pivotal moment in the evolution of crowdfunding. Online platforms began to emerge in the early 2000s, connecting creators with a broader audience beyond their immediate communities. Artist Share, launched in 2003 is often considered one of the earliest platforms that allowed musicians to fund their projects by directly engaging with their fan base (Wang 2016). This marked a shift toward a more decentralized model, leveraging the internet to connect creators and backers on a larger scale.

The mid-2000s witnessed the rise of rewards-based crowdfunding platforms such as Kickstarter and Indiegogo (Gamble et al. 2017). These platforms allowed creators to present their projects to a global audience and seek funding in exchange for non-financial rewards, such as early access to the product or exclusive merchandise. This model democratized access to funding, enabling individuals and small businesses to bring their creative ideas to life with the support of a diverse online community (Camilleri and Bresciani 2022). The next significant evolution came with the introduction of equity crowdfunding platforms in 2010. With the passage of the Jumpstart Our Business Startups (JOBS) Act in the United States in 2012, equity crowdfunding became a legal and regulated means for startups and small businesses to raise capital by offering equity or debt to a broader pool of investors (Weinberger 2020). This marked a departure from the rewards-based model, allowing backers to become actual stakeholders in the ventures they supported (Alharbey and Van Hemmen 2021). The mid-2010s saw the diversification of crowdfunding into real estate and crowdlending. Platforms like Realty Shares and Lending Club enabled investors to participate in real estate projects or lend money directly to individuals or businesses, further expanding the scope of crowdfunding beyond creative projects.

Crowdfunding has evolved into a global phenomenon, transcending geographical boundaries (Bi and Lv 2023). Platforms like GoFundMe, Kickstarter, and Crowd Cube operate on a global scale, allowing creators and entrepreneurs from different parts of the world to connect with a diverse pool of backers and investors. The proliferation of social media has played a crucial role in amplifying the reach of crowdfunding campaigns, turning them into viral phenomena (Wang and Wu 2021). The integration of blockchain technology (Patil et al., 2021) and the rise of cryptocurrency have introduced new possibilities to crowdfunding at present (Sureshbhai et al. 2020). Initial Coin Offerings (ICOs) and Security Token Offerings (STOs) leverage blockchain to enable crowdfunding for blockchain-based projects, providing backers with tokens representing ownership or other rights in the projects (Howell et al. 2020). The historical evolution of crowdfunding showcases a remarkable journey from community-based initiatives to the global, technologically-driven platforms of today. As crowdfunding continues to evolve, it reflects not only changes in finance and technology but also the democratization of investment and the power of communities to support and bring innovative ideas to fruition (Stevenson et al. 2019).

3 Types and Models of Crowdfunding

This section dissects the diverse ecosystem of crowdfunding, examining the various types and models that have proliferated (Tan and Reddy 2020). We delve into reward-based crowdfunding, where backers receive non-financial incentives, and equity crowdfunding, which offers investors a stake in the venture (Bento et al. 2019). Additionally, we explore debt crowdfunding or peer-to-peer lending, where contributors act as lenders, fueling the borrower's endeavours (Policaro 2020). By dissecting these models, readers gain a nuanced understanding of the crowdfunding landscape. It has evolved into a diverse funding landscape with various types and models, each tailored to different needs, projects, and investor preferences. Not all crowdfunding platforms are created equal. Each caters to a specific niche, offering different mechanisms for how ideas connect with support. Let us explore the four major types of crowdfunding options:

1. In rewards-based crowdfunding, backers contribute funds to a project or campaign in exchange for non-financial rewards (Wessel et al. 2021). These rewards can range from early access to the product, to exclusive merchandise, or other perks. Example Platforms: Kickstarter, Indiegogo, GoFundMe (for creative projects).
2. Equity crowdfunding allows individuals to invest in a business or startup in exchange for equity or ownership (Aggarwal et al. 2021). Investors become shareholders and stand to benefit from the success of the venture. Example Platforms: Seedrs, Crowd cube, Start Engine.
3. In debt-based crowdfunding, also known as crowdlending or peer-to-peer lending, individuals lend money to businesses or individuals in exchange for interest payments (Daskalakis and Karpouzis 2021; Gupta et al. 2023a, b). The borrower repays the loan over time. Example Platforms: Lending Club, Prosper, Funding Circle.
4. Donation-based crowdfunding involves individuals contributing funds to a cause, project, or individual without the expectation of financial returns (Jang and Chu 2022). The motivation is often philanthropic or to support a social or charitable initiative. Example Platforms: GoFundMe (for personal causes), Kickstarter (for charitable projects) (Sokolova and Perez 2018).

The above four types of crowdfunding options can also be explained through various other options such as:

- Real estate crowdfunding enables investors to pool their funds to invest in real estate projects (Borrero-Domínguez et al. 2020). This model provides an opportunity for smaller investors to participate in real estate ventures. Example Platforms: Realty Shares, Fund rise, Crowd cube (for real estate projects).
- Regulation Crowdfunding, introduced under the JOBS Act, allows startups and small businesses to raise capital from a wide pool of investors. The amount that can be raised from non-accredited investors is limited. Example Platforms: Start Engine, we funder, and Seed Invest (Wasiuzzaman 2021).
- Initial Coin Offerings (ICOs) involve raising capital for blockchain-based projects by issuing tokens or cryptocurrencies (Teuteberg and Tonnissen 2021). Investors receive tokens representing a stake or utility in the project. Example Platforms: Ethereum, Binance Launchpad (for blockchain projects).
- Similar to ICOs, Security Token Offerings (STOs) involve issuing tokens representing ownership or other rights in a project. STOs are subject to securities regulations. Example Platforms: Polymath, Securitize (Sevindik 2021).
- Royalty-based crowdfunding allows backers to receive a share of the future revenue or profits generated by a project (Pinkow 2022). This model combines elements of both rewards-based and equity-based crowdfunding. Example Platforms: Up West Labs, Koody.

- Litigation crowdfunding enables individuals or businesses to raise funds for legal expenses or litigation costs (Guy 2023). If the case is successful, backers may receive a share of the settlement. Example Platforms: Lex Shares, Crowd Justice.
- Community shares crowdfunding involves raising funds from a local community to support a specific project or business that benefits the community. Investors often become members with voting rights. Example Platforms: Crowdfunder UK, Ethex (Bone 2020).
- re-sale or pre-order crowdfunding allows businesses to gauge demand for a product by offering it for sale before production (Chu et al. 2021; Zhou et al. 2022). Backers essentially pre-purchase the product. Example Platforms: Crowd cube, Indiegogo in Demand.

The diverse types and models of crowdfunding cater to a wide range of project causes and investment preferences. Entrepreneurs, businesses, and individuals can choose the crowdfunding model that aligns with their goals (Li et al. 2018), whether they seek financial support for creative projects, business expansion, social causes, or investment opportunities (Capo and Di Pietro 2023). The ongoing evolution of crowdfunding models continues to contribute to the democratization of finance and innovation.

4 Crowdfunding Platforms—Catalysts of Change

Crowdfunding platforms serve as the linchpin of this alternative funding mechanism (Molero et al. 2020). This section provides an in-depth analysis of key crowdfunding platforms from industry giants like Kickstarter and Indiegogo to equity crowdfunding platforms such as Seedrs and Crowd Cube (Bernardino et al. 2020). This section explores their unique features, how this platform has become instrumental in driving change across various sectors, fostering innovation, supporting diverse projects and reshaping the landscape of entrepreneurship and the role it plays in connecting creators, entrepreneurs, and investors worldwide. This platform has emerged as a powerful catalyst for change, disrupting traditional finance models, innovation, and democratizing access to capital (Funk 2019).

Crowdfunding platforms have democratized access to capital by providing a direct avenue for entrepreneurs, creatives, and small businesses to secure funding. Previously, traditional financing avenues were often exclusive, favoring established entities. However, it allows individuals and businesses of all sizes to present their ideas to a global audience, attracting support from a diverse pool of backers (Oladejo et al. 2023). This democratization has proven particularly valuable for underrepresented entrepreneurs who may face challenges accessing traditional funding sources (Junge et al. 2022). Innovation thrives on diversity and experimentation and crowdfunding platforms have become fertile grounds for creative ideas to take root. This diversity of projects, fueled by the support of the crowd has led to breakthroughs in various industries, contributing to a culture of continuous innovation. It has been instrumental in channelling funds toward social causes and philanthropy. In particular, Donation-based crowdfunding enables individuals and organizations to rally support for charitable initiatives, disaster relief, medical expenses, and community projects. This democratization of philanthropy empowers ordinary people to contribute to causes they care about, creating a more inclusive and responsive approach to addressing societal challenges.

Crowdfunding platforms have also emerged as empowering tools for entrepreneurs at different stages of their journey (Srivastava et al. 2018). Early-stage startups benefit from rewards-based crowdfunding, gaining initial validation and customer support. Equity crowdfunding offers a pathway for businesses to secure investment and scale operations. For established companies, crowdfunding can serve as a means to launch new products, test markets, or engage with their customer base directly. This empowerment extends to individual creators, artists, and inventors who can bring their visions to life with the backing of a global audience (Tang et al. 2020). Investors from different corners of the globe can participate in crowdfunding campaigns, supporting projects and startups beyond their local spheres. This globalization of investment fosters cross-cultural collaboration, knowledge exchange and the democratization of economic opportunities on a global scale (Dano and Benna 2018). Crowdfunding is not just about financial transactions; it's about building communities. This community-centric approach strengthens the bond between creators and their audience, creating a network that extends beyond the funding campaign. This platform has also played a significant role in championing diversity and inclusion. By providing an open and accessible avenue for fundraising, these platforms empower entrepreneurs from diverse backgrounds, including women, minorities, and individuals in underrepresented communities (Srivastava et al. 2018). This shift toward inclusivity helps address systemic biases present in traditional funding ecosystems, promoting a more equitable distribution of resources and opportunities. As these platforms continue to evolve, they hold the potential to drive further positive transformations, shaping a future where access to capital is democratized, innovative ideas flourish, and communities actively engage in supporting the endeavours that matter to them.

5 The Impact of Crowdfunding on Entrepreneurship

Crowdfunding has democratized access to capital, particularly for entrepreneurs with innovative ideas. It has empowered the innovators and transformed funding dynamics. This section delves into the profound impact of crowdfunding on entrepreneurship, highlighting how it has empowered individuals and small businesses to turn their visions into reality and how it facilitates access to capital, drives innovation and fosters a more inclusive and dynamic startup ecosystem. It has emerged as a transformative force in the entrepreneurial landscape, reshaping the way businesses and innovators secure funding (Stroink et al. 2017).

One of the most significant impacts of crowdfunding on entrepreneurship is its role in democratizing access to capital (Ji and Robinson 2020). Previously, entrepreneurs faced significant barriers when seeking traditional funding, especially if their ideas deviated from established norms. However, it provides a direct channel for creators to showcase their projects to a global audience, enabling them to secure financial support from a diverse pool of backers. This inclusivity has been particularly beneficial for early-stage startups and those from underrepresented backgrounds. Crowdfunding allows entrepreneurs to diversify their funding sources (Sharma and Lertnuwat 2016). Instead of relying solely on traditional investors or loans, entrepreneurs can leverage crowdfunding to secure funds from a multitude of individuals, each contributing a relatively small amount. This diversified approach reduces dependence on a single source and spreads the risk among a broader base of supporters. It has become a hub for experimentation and creativity. Entrepreneurs can present innovative ideas directly to the public, bypassing traditional gatekeepers. This freedom to experiment fosters a culture of creativity, enabling entrepreneurs to pursue projects that might be considered too niche or unconventional by traditional investors (Chu et al. 2019). In particular, rewards-based crowdfunding allows entrepreneurs to validate market demand for their products or services before mass production. The direct engagement with backers provides real-time feedback and insights, helping entrepreneurs refine their offerings and tailor them to meet consumer preferences. This platform facilitates direct interaction between entrepreneurs and their backers, fostering a sense of community. This engagement extends beyond financial transactions; it creates a network of individuals who are genuinely interested in the success of the project. This community becomes a valuable asset, providing ongoing support, feedback, and word-of-mouth promotion. Backers on crowdfunding platforms often become brand advocates. Their early involvement in a project creates a sense of ownership and loyalty.

Crowdfunding reduces the dependence of small businesses and startups on traditional financial institutions (Tan and Reddy 2023). Entrepreneurs no longer need to navigate the complex and sometimes exclusive world of venture capital or banking to kick-start their projects. This independence empowers small businesses to take control of their funding journeys and make strategic decisions that align with their vision. The ability to secure funds directly from the crowd encourages entrepreneurial risk-taking. Entrepreneurs are more willing to pursue innovative and daring ideas when they have the support of a community that shares their vision. Traditional fundraising processes can be time-consuming and involve numerous intermediaries. Crowdfunding platforms streamline this process, allowing entrepreneurs to present their ideas directly to potential backers. This direct line of communication accelerates fundraising, enables entrepreneurs to access capital more quickly and bring their projects to market sooner. The iterative nature of crowdfunding campaigns enables entrepreneurs to test and iterate their ideas rapidly (Ilin et al. 2018). The real-time feedback received during a campaign allows for quick adjustments, ensuring that entrepreneurs can fine-tune their offerings based on market response. This agile approach contributes to faster product development and market entry. The impact of crowdfunding on entrepreneurship is profound, reshaping the funding landscape, fostering innovation and empowering a new generation of creators. Hence, crowdfunding has become a cornerstone of entrepreneurial success, contributing to the vibrancy and diversity of the startup ecosystem.

6 Community Engagement and Social Impact

Beyond financial transactions, crowdfunding fosters community engagement and social impact (Bone 2020). This section explores how crowdfunding creates a sense of community ownership and involvement. It also investigates the potential for crowdfunding to drive social change, support charitable causes and address societal challenges. It will highlight instances where crowdfunding has served as a force for positive impact and community-building (Simpson et al. 2021). Community engagement and social impact are integral components (Hasan 2018a, b). In terms of crowdfunding, it contributes to the broader ethos of democratized finance and collective empowerment. This platform creates a unique space for direct interaction between creators and backers, forming a supportive community. Unlike traditional funding models, crowdfunding enables entrepreneurs to engage with their audience in real-time, sharing updates, responding to queries, and building a sense of connection. This direct engagement not only builds trust but also establishes a community of individuals who are emotionally invested in the project's success. Backers contribute financially and become advocates, providing valuable insights, suggestions, and critiques that contribute to the overall improvement of the venture. They often share the project within their networks and provide valuable feedback (Hasan 2018a, b). This collaborative support network extends beyond the campaign, creating a lasting community that continues to champion the project (Ogwu et al. 2022). By presenting a project to a diverse audience, creators can gauge interest, understand potential challenges, and refine their offerings based on real-time input. This iterative process of crowdsourcing ideas ensures that projects align more closely with the preferences and needs of the intended audience. The collective wisdom of a community can significantly enhance the quality of a project.

Crowdfunding has become a powerful tool for social entrepreneurs aiming to address pressing societal challenges. Platforms dedicated to social impact and philanthropy allow individuals and organizations to present initiatives that align with social and environmental goals. Backers, driven by a shared commitment to positive change, contribute to projects that have a tangible impact on communities, ranging from sustainable development to healthcare initiatives (Bassani et al. 2019). This platform supports a diverse range of social causes, from environmental conservation to education, healthcare, and disaster relief. Creators can showcase projects that reflect their unique cultural identities, fostering a rich tapestry of creative expression. Backers, in turn, have the opportunity to support projects that contribute to a more inclusive representation of cultures and narratives. Community engagement and social impact lie at the heart of crowdfunding, transforming financial transactions into collaborative endeavours for positive change. By building supportive communities, crowd-sourcing ideas and feedback, championing philanthropy and social causes, and promoting inclusivity and diversity, crowdfunding platforms empower individuals to participate in projects that resonate with their values actively.

7 Challenges and Risks in Crowdfunding

No financial mechanism is without challenges and crowdfunding is no exception. This section addresses the risks associated with crowdfunding, exploring potential pitfalls and complexities that both creators and backers may encounter that range from campaign failures to issues of fraud and investor protection (Hossain and Oparaocha 2017a, b). By understanding these challenges, both entrepreneurs and backers can navigate the crowdfunding landscape more effectively and make informed decisions to mitigate risks (Salem et al. 2022). One of the inherent challenges in crowdfunding is the risk of failing to meet funding goals. If a campaign falls short of its target, creators may face challenges in executing their projects (Schwienbacher 2018). This failure can be attributed to various factors, including inadequate planning, ineffective marketing strategies or misjudgment of the market demand. Creators must carefully assess their funding requirements and set realistic goals to mitigate the risk of compromised project viability. For backers, the risk lies in contributing to a project that does not reach its funding goal. In such cases, backers may not receive the promised rewards or benefits, leading to disappointment and potential distrust in the crowdfunding model. Clear communication from creators about the potential risks and challenges, along with transparent updates throughout the campaign, can help manage backer expectations in the event of funding shortfalls. Creators may misrepresent their projects or intentions, leading to financial loss for backers who may not receive the promised returns or rewards. Instances of fraud or scams can tarnish the reputation of crowdfunding platforms. Platforms must implement robust vetting processes to ensure the legitimacy of projects (Chen 2023). Likewise, backers should exercise due diligence before contributing to campaigns, researching creators and evaluating the feasibility of the project.

Creators may face challenges related to the protection of their intellectual property. Crowdfunding exposes ideas and designs to a wide audience, potentially increasing the risk of imitation or unauthorized use (Rei-Anderson 2022). Navigating the balance between showcasing a project to attract backers and safeguarding intellectual property can be a complex challenge. Backers may face the risk of supporting projects that infringe on existing intellectual property rights. This can result in legal challenges, including potential lawsuits against creators. Clear communication between creators and backers regarding intellectual property rights and potential risks is crucial in addressing these concerns. For creators, delivering on the promises made during a crowdfunding campaign can be challenging. Unforeseen obstacles, such as production delays, manufacturing issues, or supply chain disruptions, can impact project timelines. Managing backer expectations and providing regular updates become essential in mitigating risks associated with project execution. Backers face the risk of not receiving the promised rewards or experiencing delays in project delivery. This can lead to dissatisfaction and a loss of trust in crowdfunding as a reliable funding model. Creators must prioritize transparent communication and actively address any challenges that may arise during the execution phase. Creators and platforms must stay abreast of changes in regulations to ensure compliance. Failure to adhere to regulatory requirements can lead to legal consequences, affecting the viability of projects and the reputation of crowdfunding platforms (Rustem et al. 2021). Understanding the regulatory framework and choosing platforms that prioritize compliance is crucial for both creators and backers.

From navigating funding goals and potential fraud to managing intellectual property concerns and ensuring regulatory compliance, stakeholders must approach crowdfunding with careful consideration, transparency and a commitment to building trust within the community. Proactive communication, due diligence and a clear understanding of the risks can contribute to a more resilient and sustainable crowdfunding landscape. As we conclude this exploration of crowdfunding as an alternative funding option, we reflect on its transformative power. It has become a dynamic force shaping the future of finance, unlocking opportunities, fostering innovation, and building communities around shared passions. Crowdfunding has emerged as a transformative force, reshaping the future of finance by introducing a paradigm shift in how capital is raised, invested and distributed. It has democratized access to capital, dismantling traditional barriers that often-excluded small businesses, startups, and unconventional ideas. Entrepreneurs from diverse backgrounds now have a direct avenue to present their projects to a global audience, attracting funding from individuals who resonate with their vision. This democratization ensures that funding opportunities are not limited to a select few, promoting a more inclusive and diverse entrepreneurial ecosystem. Crowdfunding enables individual investors to participate in ventures that align with their interests and values. Previously, such investment opportunities were often reserved for institutional investors or high-net-worth individuals (Hasan et al. 2023). Through crowdfunding, individuals of varying financial capacities can become stakeholders in projects ranging from creative endeavours to innovative startups, transforming them from passive consumers to active contributors in the financial landscape. It allows entrepreneurs to validate their concepts before full-scale production. The direct interaction with backers provides valuable market insights, helping creators refine their offerings based on real-time feedback. This iterative process of validation ensures that products and services are not only innovative but also tailored to meet the evolving needs and preferences of the market.

Crowdfunding transcends geographical barriers, connecting entrepreneurs with backers from around the world (Nunzia 2020). This global connectivity ensures that innovative ideas and projects are not confined to specific regions or markets. Entrepreneurs can leverage the diverse perspectives and support from a global community, fostering cross-cultural collaboration and expanding the reach of their ventures. It contributes to financial inclusion by providing opportunities for individuals who may have limited access to traditional financial services. Whether supporting a charitable cause, investing in a startup, or contributing to a creative project, individuals from various socioeconomic backgrounds can actively participate in the financial ecosystem. This inclusivity empowers a broader segment of the population to engage in the dynamics of finance, promoting economic diversity. By connecting creators directly with backers, the need for centralized financial institutions is diminished. This shift towards decentralization promotes a more transparent and peer-to-peer approach, fostering trust and accountability within the crowdfunding community. The peer-to-peer nature of crowdfunding aligns with the principles of decentralization. This empowerment of peer-to-peer transactions challenges the conventional models of finance, encouraging a more distributed and collaborative approach to economic activities. Crowdfunding redefines the dynamics between investors and entrepreneurs by encouraging active engagement and collaboration. This shift in dynamics fosters a sense of community, where backers are emotionally invested in the ventures they support, creating a symbiotic relationship that extends beyond the financial transaction. Unlike traditional financing where the burden of risk falls heavily on the entrepreneur, crowdfunding distributes risk among a larger pool of backers. Similarly, the success of a project translates into shared rewards for backers, creating a more equitable and collaborative approach to entrepreneurship and investment.

8 Implications and Strategies for Effective Growth of Crowdfunding

Here are the key implications and strategies to leverage the power of community on your crowdfunding journey for entrepreneurial growth:

- **Pre-launch buzz:** Do not hit the launch button without a bang. Generate pre-campaign excitement through social media, email marketing, and influencer partnerships. Create a landing page, share teasers, and build anticipation to make your launch day a success.
- **Storytelling that resonates:** People connect with emotion. Share your vision, the problem you are solving, and the impact you will make with genuine passion. Let your personality shine through, showcase your team's expertise, and weave a narrative that draws the crowd in.
- **Rewards that excite:** Your rewards are your thank-you and incentives. Make them unique, relevant to your project, and something your backers will truly value. Offer exclusive experiences, personalized gifts, or early access to your product. Remember, it is not just about the monetary value but the emotional connection.
- **Transparency and updates:** Your backers are your partners. Keep them informed about progress, challenges, and milestones. Share behind-the-scenes glimpses, answer their questions promptly, and address any concerns with honesty and humility. This builds trust and strengthens the bond that fuels your campaign.

- **Celebrating every win:** Every contribution, big or small, is a victory. Acknowledge your backers, express your gratitude, and celebrate each achievement along the way. Share milestones, host virtual events, and make your backers feel like an integral part of your success story.

As this vibrant ecosystem continues to evolve, new models will emerge, technology will enhance the experience and the reach of crowdfunding will extend even further

9 Conclusion

Crowdfunding's role in shaping the future of finance is characterized by its transformative impact on democratization, innovation, global connectivity, decentralization and the redefinition of investor-entrepreneur dynamics. As a dynamic and inclusive funding model, crowdfunding continues to break traditional moulds, paving the way for a financial landscape where diverse voices, ideas, and contributors collectively drive positive change and shape the future of finance in unprecedented ways. It is a testament to the collective power of belief, the desire to be part of something bigger, and the human spirit's unwavering pursuit of innovation.

References

Aggarwal R, Lee MJ, Osting B, Singh H (2021) Improving funding operations of equity-based crowdfunding platforms. Prod Oper Manag 30(11):4121–4139. https://doi.org/10.1111/poms.13505

Alharbey M, Van Hemmen S (2021) Investor intention in equity crowdfunding. Does Trust Matter? J Risk and Finan Managem 14(2). https://doi.org/10.3390/jrfm14020053

Baber H, Fanea-Ivanovici M (2022) Fifteen years of crowdfunding–a bibliometric analysis. Technology analysis and strategic management. pp 1–15

Baber H, Fanea-Ivanovici M (2023) Motivations behind backers' contributions in reward-based crowdfunding for movies and web series. Int J Emerg Mark 18(3):666–684

Bassani G, Marinelli N, Vismara S (2019) Crowdfunding in healthcare. J Technol Transfer 44(4):1290–1310. https://doi.org/10.1007/s10961-018-9663-7

Baumgardner T, Neufeld C, Huang PCT, Sondhi T, Carlos F, Talha MA (2017) Crowdfunding as a fast-expanding market for the creation of capital and shared value. Thunderbird Int Business Rev 59(1):115–126

Belleflamme P, Omrani N, Peitz M (2015) The economics of crowdfunding platforms. Inf Econ Policy 33:11–28

Bento N, Gianfrate G, Thoni MH (2019) Crowdfunding for sustainability ventures. J Cleaner Prod 237. https://doi.org/10.1016/j.jclepro.2019.117751

Bernardino S, Santos JF, Oliveira S (2020) Financing nascent entrepreneurs by rewardbased crowdfunding: lessons from indiegogo campaigns. In: Handbook of research on nascent entrepreneurship and creating new ventures. IGI Global. https://doi.org/10.4018/978-1-7998-4826-4.ch011

Bi G, Lv J (2023) Can an early exit mechanism attract more pledges in equity-based crowdfunding? Evidence from China. Finan Res Lett 56. https://doi.org/10.1016/j.frl.2023.104166

Bone J (2020) Crowdfunding social or community-focused projects. In: CrowdAsset: crowdfunding for policymakers. World Scientific Publishing Co. https://doi.org/10.1142/9789811207822_0008

Borrero-Domínguez C, Cordón-Lagares E, Hernández-Garrido R (2020) Sustainability and real estate crowdfunding: success factors. Sustain (Switzerland) 12(12). https://doi.org/10.3390/su12125136

Camilleri MA, Bresciani S (2022) Crowdfunding small businesses and startups: a systematic review, an appraisal of theoretical insights and future research directions. Eur J Innov Manag. https://doi.org/10.1108/EJIM-02-2022-0060

Capo F, Di Pietro F (2023) Entrepreneurial finance, crowdfunding, and language: from social to financial support. In: Entrepreneurial finance, crowdfunding, and language: from social to financial support. Taylor and Francis. https://doi.org/10.4324/9781003349860

Chen WD (2023) Crowdfunding: different types of legitimacy. Small Bus Econ 60(1):245–263. https://doi.org/10.1007/s11187-022-00647-0

Chu C-C, Cheng Y-F, Tsai F-S, Tsai S-B, Lu K-H (2019) Open innovation in crowdfunding context: diversity, knowledge, and networks. Sustainability (Switzerland) 11(1). https://doi.org/10.3390/su11010180

Chu T, Wei X, Zhou Y (2021) The pricing and efficiency of pre-Sale crowdfunding. Finan Res Lett 40. https://doi.org/10.1016/j.frl.2020.101793

Cumming D, Groh AP (2018) Entrepreneurial finance: unifying themes and future directions. J Corp Finan 50:538–555. https://doi.org/10.1016/j.jcorpfin.2018.01.011

Daniele LM (2021) Crowdfunding and socially responsible initiatives. In: The evolution of sustainable investments and finance: theoretical perspectives and new challenges. Springer International Publishing. https://doi.org/10.1007/978-3-030-70350-9_4

Dano UL, Benna UG (2018) Africa and Indian crowdfunding markets: a cross-cultural comparative analysis. In: Crowdfunding and sustainable urban development in emerging economies. IGI Global. https://doi.org/10.4018/978-1-5225-3952-0.ch007

Daskalakis N, Karpouzis E (2021) Exploring determinants in cross-border activity in equity crowdfunding and peer-to-peer lending, from a user's perspective. Small Enterp Res 28(3):293–313. https://doi.org/10.1080/13215906.2021.1989624

Doe-Bruce O (2019) Blockchain and alternative sources of financing. Cryptofinance and mechanisms of exchange: the making of virtual currency. pp 91–111

Fanea-Ivanovici M, Baber H (2021a) The role of entrepreneurial intentions, perceived risk and perceived trust in crowdfunding intentions. Eng Econ 32(5):433–445

Fanea-Ivanovici M, Baber H (2021b) Predicting entrepreneurial and crowdfunding intentions-a study of Romania and South Korea. Amfiteatru Econ 23(15):1003–1014

Funk AS (2019) Traditional informal finance versus crowdfunding in China. In: Contributions to management science, pp 199–214. https://doi.org/10.1007/978-3-319-97253-4_6

Galuszka P, Brzozowska B (2017) Crowdfunding and the democratization of the music market. Media Cult Soc 39(6):833–849. https://doi.org/10.1177/0163443716674364

Gamble JR, Brennan M, McAdam R (2017) A rewarding experience? Exploring how crowdfunding is affecting music industry business models. J Bus Res 70:25–36. https://doi.org/10.1016/J.JBUSRES.2016.07.009

Gao Q (2016) Empirical studies of online crowdfunding

Gerber EM, Hui JS, Kuo PY (2012) Crowdfunding: why people are motivated to post and fund projects on crowdfunding platforms. In: Proceedings of the international workshop on design, influence, and social technologies: techniques, impacts and ethics, February, vol 2(11). pp 10

Gupta SK, Tiwari S, Hassan A, Gupta P (2023) Moderating effect of technologies into behavioural intensions of tourists toward use of mobile wallets for digital payments: TAM model perspective. Int J Hospitality and Tourism Syst 16(1):43–57

Gupta D, Singhal A, Sharma S, Hasan A, Raghuwanshi S (2023) Humans' emotional and mental well-being under the influence of artificial intelligence. J ReAttach Therapy and Developm Diversities 6(6s):184–197

Guy S (2023) Mobilising the market: an empirical analysis of crowdfunding for judicial review litigation. Mod Law Rev 86(2):331–363

Hasan A (2018) Impact of store and product attributes on purchase intentions: an analytical study of apparel shoppers in Indian organized retail stores. Vision 22(1):32–49

Hasan A (2018) Evaluation of factors influencing exclusive brand store choice: an investigation in the Indian retail sector. Vision 22(4):416–424

Hasan MB, Hassan MK, Alhomaidi A (2023) How do sectoral Islamic equity markets react to geopolitical risk, economic policy uncertainty, and oil price shocks?. J Econ Asymmetries 28:e00333

Hervé F, Schwienbacher A (2019) Crowdfunding and innovation. In: Contemporary topics in finance: a collection of literature surveys. pp 331–349

Hommerová D (2020) 13 crowdfunding as a new model of nonprofit funding. Financing Nonprofit Organizations, pp 173

Hossain M, Oparaocha GO (2017) Crowdfunding: motives, definitions, typology and ethical challenges. Entrepreneurship Res J 7(2). https://doi.org/10.1515/erj-2015-0045

Hossain M, Oparaocha GO (2017) Crowdfunding: motives, definitions, typology and ethical challenges. Entrepreneurship Res J 7(2)

Howell ST, Niessner M, Yermack D (2020) Initial coin offerings: financing growth with cryptocurrency token sales. Rev Finan Stud 33(9):3925–3974. https://doi.org/10.1093/rfs/hhz131

Huang S, Pickernell D, Battisti M, Nguyen T (2022) Signalling entrepreneurs' credibility and project quality for crowdfunding success: cases from the Kickstarter and Indiegogo environments. Small Business Econ 1–21

Ilin P, Platis D, Hammouda I (2018) Insights into the trilateral relationship of crowdfunding campaigns, open source and communities. In: IFIP advances in information and communication technology, vol 525. pp 26–36. https://doi.org/10.1007/978-3-319-92375-8_3

Inbar Y, Barzilay O (2014) Community impact on crowdfunding performance. Available at SSRN 2524910

Jang H, Chu W (2022) The effect of message features on donations in donation-based crowdfunding. J Consum Behav 21(6):1464–1477. https://doi.org/10.1002/cb.2099

Ji M, Robinson T (2020) Democratizing capital: building union coop partnerships through economically targeted investing and crowdfunding innovations. In: Cooperatives and social innovation: experiences from the Asia pacific region. Springer, Singapore. https://doi.org/10.1007/978-981-15-8880-8_17

Junge LB, Laursen IC, Nielsen KR (2022) Choosing crowdfunding: why do entrepreneurs choose to engage in crowdfunding? Technovation 111. https://doi.org/10.1016/j.technovation.2021.102385

Koch JA, Siering M (2015) Crowdfunding success factors: the characteristics of successfully funded projects on crowdfunding platforms

Lee CH, Chiravuri A (2019) Dealing with initial success versus failure in crowdfunding market: serial crowdfunding, changing strategies, and funding performance. Internet Res 29(5):1190–1212. https://doi.org/10.1108/INTR-03-2018-0132

Li H, Chen X, Zhang Y, Hai M (2018) Prediction of financing goal of crowdfunding projects. Proc Comput Sci 139:108–113. https://doi.org/10.1016/j.procs.2018.10.225

Molero NL, Gómez EH, Briones Peñalver AJ (2020) Crowdfunding. The alternative to traditional financing of social projects; [Crowdfunding Social. La alternativa a la financiación tradicional de proyectos de organizaciones de participación]. REVESCO Revista de Estudios Cooperativos 136. https://doi.org/10.5209/reve.71851

Murray A, Kotha S, Fisher G (2020) Community-based resource mobilization: how entrepreneurs acquire resources from distributed non-professionals via crowdfunding. Organ Sci 31(4):960–989

Nunzia C (2020) The role of geographical clusters in the success of reward-based crowdfunding campaigns. Int J Entrepreneurship and Innov 21(4):250–262. https://doi.org/10.1177/1465750320915915

Ogwu K, Hickey P, Okeke OJ-P, Haque A, Pimenidis E, Kozlovski E (2022) An integrated online/offline social network-based model for crowdfunding support in developing countries: the case of Nigeria. Sustainability (Switzerland) 14(15). https://doi.org/10.3390/su14159333

Oladejo OR, Ogunsade AI, Kolade S, Atiase V (2023) Crowdfunding and traditional finance: the prospects and challenges for SMEs in Nigeria. In: Entrepreneurial financial resilience and financial innovation in a turbulent era. Edward Elgar Publishing Ltd. https://www.scopus.com/inward/record.uri?eid=2-s2.0-85167697101&partnerID=40&md5=e88052147d8212f6c011476cf5fc6ddb

Parhankangas A, Mason C, Landström H (2019) Crowdfunding: an introduction. In: Handbook of research on crowdfunding. pp 1–21

Patil V, Gupta V, Sarode R (2021) Blockchain-based crowdfunding application. In: Proceedings of the 5th international conference on I-SMAC (IoT in social, mobile, analytics and cloud), I-SMAC 2021, pp 1546–1553. https://doi.org/10.1109/I-SMAC52330.2021.9640888

Pinkow F (2022) The impact of common success factors on overfunding in reward-based crowdfunding: an explorative study and avenues for future research. J Entrepreneurship, Managem Innov 18(1):131–167. https://doi.org/10.7341/20221815

Policaro GA (2020) Debt crowdfunding. a new financing model for smes in Italy; [Il debt crowdfunding: Un nuovo modello di finanziamento per le PMI in Italia]. Osservatorio del Diritto Civile e Commerciale 9(1):161–188. https://doi.org/10.4478/98135

Rei-Anderson C (2022) Copyright, podcasts and crowdfunding: an 'abundance model' for an emerging medium? Queen Mary J Intellect Property 12(2):185–205. https://doi.org/10.4337/qmjip.2022.02.02

Rustem M, Natalya Y, McKeown T (2021) Legal regulation of crowdfunding. In: Proceedings—international conference on developments in Esystems engineering, DeSE, 2021-December, pp 599–603. https://doi.org/10.1109/DESE54285.2021.9719475

Ryu S, Suh A (2021) Online service or virtual community? Building platform loyalty in reward-based crowdfunding. Internet Res 31(1):315–340

Salem IE, Elkhwesky Z, Baber H, Radwan M (2022) Hotel resuscitation by reward-based crowdfunding: a critical review and moderated mediation model. Tourism Managem Perspect 43:100986

Schwienbacher A (2018) Entrepreneurial risk-taking in crowdfunding campaigns. Small Bus Econ 51(4):843–859. https://doi.org/10.1007/s11187-017-9965-4

Sevindik V (2021) Blockchain based resource tokenization for crowdfunding of wireless network investment. In: 2021 11th IFIP international conference on new technologies, mobility and security, NTMS 2021. https://doi.org/10.1109/NTMS49979.2021.9432653

Sharma S, Nilubol Lertnuwat A (2016) The financial crowdfunding with diverse business models. J Asian and African Soc Sci Humanit 2(2):74–89. https://www.aarcentre.com/ojs3/index.php/jaash/article/view/66/271

Shneor R (2020) Crowdfunding models, strategies, and choices between them. In: Advances in crowdfunding: research and practice. pp 21–42

Simpson B, Schreier M, Bitterl S, White K (2021) Making the world a better place: how crowdfunding increases consumer demand for social-good products. J Mark Res 58(2):363–376. https://doi.org/10.1177/0022243720970445

Sokolova K, Perez C (2018) The digital ingredients of donation-based crowdfunding. A data-driven study of Leetchi projects and social campaigns. J Decis Syst 27(3):146–186. https://doi.org/10.1080/12460125.2019.1587133

Srivastava S, Oo PP, Sahaym A, Allison TH (2018) Kickstart or jumpstart? Understanding women entrepreneurs' crowdfunding performance. In: A research agenda for women and entrepreneurship: identity through aspirations, behaviors and confidence. Edward Elgar

Publishing Ltd. https://www.scopus.com/inward/record.uri?eid=2-s2.0-85075492724&partnerID=40&md5=ef30188dbac88e30e521fbacf0befaf9
Stemler AR (2013) The JOBS act and crowdfunding: harnessing the power—and money—of the masses. Bus Horiz 56(3):271–275
Stevenson RM, Kuratko DF, Eutsler J (2019) Unleashing main street entrepreneurship: crowdfunding, venture capital, and the democratization of new venture investments. Small Bus Econ 52(2):375–393. https://doi.org/10.1007/s11187-018-0097-2
Stroink M, Nelson CH, Davis AC (2017) Using a complexity lens to address local food dilemmas in Northern Ontario: the viability of crowdsourcing and crowdfunding. In: Nourishing communities: from fractured food systems to transformative pathways. Springer International Publishing. https://doi.org/10.1007/978-3-319-57000-6_7
Sureshbhai PN, Bhattacharya P, Tanwar S (2020) KaRuNa: a blockchain-based sentiment analysis framework for fraud cryptocurrency schemes. In: 2020 IEEE international conference on communications workshops, ICC Workshops 2020—proceedings. https://doi.org/10.1109/ICCWorkshops49005.2020.9145151
Tan YH, Reddy SK (2023) Sustainable funding for small businesses: an investigation into the dynamics of the recurring crowdfunding model. Int Small Business J: Res Entrepreneurship. https://doi.org/10.1177/02662426231205758
Tan YH, Reddy SK (2020) Crowdfunding platforms: ecosystem and evolution. Foundat Trends in Market 14(2):53–172. https://doi.org/10.1561/1700000061
Tang L, Baker R, An L (2020) The success of crowdfunding projects: technology, globalization, and geographic distance. Econ Innov New Technol. https://doi.org/10.1080/10438599.2020.1838412
Teuteberg F, Tonnissen S (2021) Success factors of crowdfunding with initial coin offerings-—An exploratory analysis; [Erfolgsfaktoren von Crowdfunding mit Initial Coin Offerings-eine explorative Analyse]. Betriebswirtschaftliche Forschung und Praxis 2021(3):302–322. https://www.scopus.com/inward/record.uri?eid=2-s2.0-85114357037&partnerID=40&md5=78ce2a5a134fc341b64f7c88688a7263
Wachira EW, Wachira VK (2021) A crowdfunding Kenyában: A sikeres kampány tényezői: a Kickstarter crowdfunding-platform esete. Public Finance Quarterly 66(3):423–439. https://doi.org/10.35551/PSZ_2021_3_6
Wang W, Wu YJ (2021) Online financing campaigns' comments: insights from crowdfunding pitches. In: VA, LMD, ANR (eds) Springer proceedings in complexity. Springer Science and Business Media B.V., pp 485–493. https://doi.org/10.1007/978-3-030-62066-0_37
Wang C (2016) The promise of kickstarter: extents to which social networks enable alternate avenues of economic viability for independent musicians through crowdfunding. Soc Media and Soc 2(3). https://doi.org/10.1177/2056305116662394/ASSET/IMAGES/LARGE/10.1177_2056305116662394-FIG6.JPEG
Wasiuzzaman S (2021) Regulations, perceived information quality and perceived risk of equity crowdfunding: a study of Malaysian investors. Strateg Chang 30(4):353–366. https://doi.org/10.1002/jsc.2429
Weinberger DJ (2020) Civic crowdfunding, equity, and the role of government. In: CrowdAsset: crowdfunding for policymakers. World Scientific Publishing Co. https://doi.org/10.1142/9789811207822_0003
Wessel M, Gleasure R, Kauffman RJ (2021) Creators and backers in rewards-based crowdfunding: will incentive misalignment affect Kickstarter's sustainability? In: BTX (ed) Proceedings of the annual hawaii international conference on system sciences (Vols 2020-January). IEEE Computer Society, pp 6543–6552. https://www.scopus.com/inward/record.uri?eid=2-s2.0-85108375594&partnerID=40&md5=6b86437ea1614bc5c372201fa749819f

Zhang H, Chen W (2019) Crowdfunding technological innovations: interaction between consumer benefits and rewards. Technovation 84–85:11–20. https://doi.org/10.1016/j.technovation.2018.05.001

Zhou Y, Cui J, Wang N (2022) Do backers prefer crowdfunding or pre-order? An empirical study. Front Psychol 13. https://doi.org/10.3389/fpsyg.2022.984775

Medical Crowdfunding- "We Make a Living by What We Get but We Make a Life by What We Give"

Mina Fanea-Ivanovici and Hasnan Baber

Abstract The primary objective of the chapter is to delve into the intricacies of medical crowdfunding and to comprehend the factors that drive individuals to contribute to medical crowdfunding campaigns. By exploring the dynamics of this emerging method of fundraising we seek to shed light on the motivations that prompt individuals to support such important causes. Understanding the underlying incentives behind donating to medical crowdfunding campaigns can provide valuable insights into the altruistic behaviors and social factors that fuel this impactful avenue of financial support for individuals in need of medical assistance. The study will be helpful for people living in countries with no national insurance schemes as they can use crowdfunding to pledge funds globally. The study will also be helpful for researchers and crowdfunding platforms to understand the nature and factors that influence people to donate to such benevolent causes.

Keywords Medical crowdfunding · Healthcare · Empathy · Altruism · Social presence

1 Introduction

The recent COVID-19 pandemic has exposed the healthcare system of most developed countries and the situation in developing countries is even worse. Developed countries like the US UK Italy France and Germany are among the 10 nations with the highest reported death rates during the COVID-19 pandemic. On the other hand

M. Fanea-Ivanovici (✉)
Department of Economics and Economic Policies, Bucharest University of Economic Studies, 6 Piata Romana District 1, 010374 Bucharest, Romania
e-mail: mina.ivanovici@economie.ase.ro

H. Baber
Department of Management, Strategy and Entrepreneurship, School of Business Administration, American University of Sharjah, 26666, Sharjah, United Arab Emirates
e-mail: hbaber@aus.edu

H. Baber and M. Fanea-Ivanovici (eds.), *Sustainable Financing—A Contemporary Guide for Green Finance, Crowdfunding and Digital Currencies*, World Sustainability Series, https://doi.org/10.1007/978-3-031-80969-9_9

developing countries e.g. India Brazil Mexico and Peru share the top of the 10 nations having the highest reported number of deaths due to COVID-19. Developing countries particularly South Asian countries struggled like developed countries or even worse due to a lack of healthcare infrastructure and policies to provide medical assistance to infected patients. In India a family in Kolkata was charged a bill of 1.5 million rupees ($20,400) for the treatment of their 57-year-old mother. However a third of the amount was settled later and paid by the family (Thiagarajan 2020). The situation in densely populated countries like Bangladesh and Pakistan was even worse (Waris et al. 2020). A study by Abid et al. (2020) reported that Pakistan did not have adequate means to deliver to patients during the peak of COVID-19. The economic conditions of the people added to the anxiety stress and mental health of the citizens of Bangladesh during COVID-19 (Didar-Ul Islam et al. 2020).

Besiders the misery of the people soaring medical bills left people to pay or pledge help from the public. Healthcare systems in developing countries have little capacity to handle large patient volumes as many people do not have health insurance coverage thereby living at the risk of disastrous health expenditures (Loayza and Pennings 2020). During the COVID-19 pandemic it was for the first time that a mammoth organization like the World Health Organization (WHO) requested people to donate financially towards healthcare. Pledging funds from large populations usually small amounts through an online platform is called 'crowdfunding'. The biggest advantage of such a fundraising process is global reach. GoFundMe.com a crowdfunding platform had raised funds for around 150,000 COVID-19-related campaigns till 31 August 2020 (Snyder et al. 2022).

Crowdfunding is a means of raising funds through online platforms and has four types –equity-based lending-based reward-based and donation-based crowdfunding (Baber and Fanea-Ivanovici 2023). Donation-based crowdfunding is a volunteer deed where the supporter does not get any intangible or physical return. Reward-based crowdfunding pledges funds from individuals and organizations in return for the delivery of product produced (sometimes customized) (Vismara 2019). Crowdfunding particularly donation-based has increasingly received an overwhelming response from donors for benevolent causes e.g. medical treatments financial hardships of families and regions facing natural catastrophes.

Recently individuals and nonprofit organizations have used crowdfunding particularly donation-based for healthcare. This does not only involve financing individual severe medical cases but also the increased need for funding health initiatives e.g. the development of vaccines or drugs for public health. This alternate financing channel even raises funds for improving research protocols and systems and has given new life to the health sector (Ren et al. 2020). Medical crowdfunding must have an emphatic touch and inculcate the altruistic side of the campaign. For example, a picture showing a kid who needs financial help to cover medical costs is the most effective way to generate funds. Also, a picture showing a severe medical health condition generates more funds than relative normalcy in the condition (Ren et al. 2020). There is much need for research on donation-based medical crowdfunding projects which heavily depend on empathy trust and altruism.

This chapter will be useful for researchers crowdfunding platforms fundraising organizations and non-profit organizations to understand the nature of donors and how and which various factors contribute to forming intentions. The conclusions will open the scope for future studies to investigate the intentions of donors at the time of natural catastrophes e.g. earthquakes military conflicts such as the ongoing war in Ukraine and Gaza medical emergencies or crises pandemics or endemic mass destruction and so on.

2 Background

The existing literature has delved into medical crowdfunding as an emerging phenomenon both by looking at its macroeconomic determinants (from a demand-side perspective of funds) and the microeconomic determinants of donation (from a supply-side perspective of funds). From a demand-side perspective, the research conducted by Bassani et al. (2019) explores the reasons for the healthcare crowdfunding emergence and where it mostly occurs based on the analysis of a worldwide population of such platforms. This study relates medical crowdfunding to national healthcare systems and finds evidence of a substitution effect in the case of low public coverage. However it is not only emerging or developing economies where people resort to medical crowdfunding. Developed countries such as the United States use medical crowdfunding due to the shortcomings of the healthcare and health insurance systems (Berliner and Kenworthy 2017).

Successful campaigns depend on medical and media literacies which may at the same time have the opposite effect of further generating unintended inequities in society and of how pooled resources are distributed among those in need (Kubheka 2020). In other words the most skillful in marketing will attract more resources whereas the major issue of incapable public policy to address health issues lingers and gives rise to more medical crowdfunding campaigns. Research conducted on the United States' medical crowdfunding reveals that between 2016 and 2020 as few as 12% of the campaigns met their financial goal with the lowest funds raised in areas where people have more medical debt in places with higher rates of uninsured and lower incomes (Kenworthy and Igra 2022). These findings highlight the further gaps created in society in terms of opportunities to tap the donations made by the public. That is why certain authors plead for regulation in medical crowdfunding to render the medical crowdfunding market more stable and efficient and to achieve its ultimate goal of improving public health (Renwick and Mossialos 2017). Successful campaigns are achieved by the digitally literate those who are already part of social networks and have exposure and those who are emotionally potent (Kubheka 2020).

From the supply-side perspective the act of donation to a charity or prosocial cause is triggered by personal factors such as impure altruism that gives rise to intrinsic motivations (the so-called warm-glow effect) and factors such as joy belongingness and satisfaction. This is an important aspect when considering medical crowdfunding especially because the utilitarian value is harder to satisfy given asymmetric information and moral hazard and because it is expected that no hedonic value stems from medical projects (Zhao and Vinig 2017). Other researchers suggest that the intention to donate can be explained by the theory of persuasion i.e. the effects of rational emotional and credibility appeals on donors (Wu et al. 2022). Rational and credible appeals in the message (Majumdar and Bose 2018) on the one hand and emotional impression of the call for finance (Kim et al. 2022) are determinants of successful donation-based crowdfunding campaigns. Shneor et al. (2023) find that when there are pre-existing relations between fundraisers and backers contributions are higher. Similarly networks are critical to achieving crowdfunding success according to Zhao and Vinig (2020). The chances of attracting funding may also depend on the level of self-finance provided by the project initiator (Zhao et al. 2022). Higher donations for medical projects are more likely to occur on non-investment-based or not exclusively healthcare platforms according to Bassani et al. (2019) therefore the odds of success also depend on the virtual market where the demand and supply of funds meet.

3 Driving Forces to Donate in Medical Crowdfunding Campaigns

3.1 Theory of Planned Behaviour

The theory of planned behavior is a widely used model to check the intentions of an individual. It has mainly three factors that shape the intention- self-efficacy attitude and social norm. Self-efficacy refers to the perceived capacity or capability to complete a task and it is "concerned not with what one has but with belief in what one can do with whatever resources one can muster" (Bandura 2007; 646). Another definition of self-efficacy is performance expectancy in completing a task (Kuo and Wu 2014). Self-efficacy and perceived behavioral control have been used interchangeably with the former being preferred in the crowdfunding literature (Kuo and Wu 2014; Shneor and Munim 2019). A large body of literature confirmed that there is a positive and significant influence of self-efficacy on behavioral intentions in general (Huang et al. 2016; Judge et al. 2007) and particularly in crowdfunding (Baber 2022; Baber et al. 2022; Kuo and Wu 2014; Shneor and Munim 2019).

For the specific case of medical crowdfunding research on the Chinese users of mobile social networking sites indicates that there is a positive correlation between perceived behavioral control and people's donation intention (Liu et al. 2021a). Similarly self-efficacy and crowdfunding intentions for charitable campaigns were identified to be a positive and significant relationship due to extrinsic and intrinsic motivations (Chen et al. 2021). Another relevant Asian study on donation behavior conducted in Oman confirms the positive and relevant relationship between perceived behavioral control and intention to donate in crowdfunding campaigns (Allah Pitchay et al. 2022). Data on South Korean crowdfunders indicate that crowdfunding behavioral intentions for sustainable development significantly and positively depend on perceived behavioral control (Kim and Petrick 2021) and the same is checked for Korean crowdfunding projects implying co-creation (Kim and Hall 2019).

Within the theory of planned behavior attitude was defined as an individual's evaluation of a subject matter (Ajzen and Fishbein 1977). The authors posit that attitude has a central place in shaping behavior and that a strong relationship between attitude and behavior depends on a high correspondence between the target and action elements. A study on the Spanish population revealed a strong and positive correlation between attitudes and intentions to donate to crowdfunding campaigns for rural development projects (Pérez y Pérez and Egea 2019). The correlation was later tested on countries having opposing cultures (i.e. individualism and collectivism)—Finland and China in an extended Theory of Planned Behavior framework by analyzing the answers provided by users of national platforms and the results were in line with the initial theory (Shneor et al. 2021). For China the positive and strong correlation was also confirmed in more recent studies in the realm of medical crowdfunding projects that are posted on mobile social networking sites (Liu et al. 2021b). Numerous other studies discuss the correlation between attitude and intention of participation in crowdfunding for diverse geographical regions (Asia Europe Africa) in various other areas such as entrepreneurship (attitude—entrepreneurial intention) (Baber 2022; Fanea-Ivanovici and Baber 2021a b) crowdfunding for the hospitality industry (Baber et al. 2022).

Part of the Theory of Planned Behavior subjective norms as described in the literature refer to one's belief that the close ones such as important people in their life or the social group they are part of will approve (or not) of their actions. In other words subjective norms answer the question: 'Do other people want me to do that?' (Ham et al. 2015). Subjective norms fall into two categories i.e. social norms and descriptive norms (Ham et al. 2015) and have been credited with the capacity to influence behavior (Venkatesh et al. 2012). Subjective norms in crowdfunding can be described as the pressure that an individual may perceive from others regarding their participation as a funder in crowdfunding campaigns (Cecere et al. 2017; Moon and Hwang 2018). Most studies analyzing the influence of subjective norms on crowdfunding intention indicate a positive and significant correlation between the two variables. (Moon and Hwang 2018; Pan and Liu 2018; Shneor and Munim 2019) with just a few exceptions (Chen et al. 2019).

3.2 Empathy

Empathy which is "an affective state that stems from the apprehension of another's emotional state or condition" has been investigated in the context of prosocial or philanthropy intentions and behavior (Eisenberg and Miller 1987). Empathy for the recipients of donations was found to be significant in explaining donation behavior for various prosocial causes (Lee et al. 2014). Moreover empathy as an emotional mechanism also plays an important role in customers when they respond to pro-social products and pro-social marketing and as a result they are willing to pay more for such products (Lee 2016).

Both empathy and sympathy are emotional factors that trigger donation behavioral intentions on pro-social crowdfunding platforms in the absence of any type of reward or benefit for the donor (Rhue and Robert 2018). Therefore campaign managers normally exploit the emotional touch in order to achieve financial targets. In charitable crowdfunding empathy and perceived credibility weigh greatly in an individual's intention to contribute to such projects and project popularity and empathy are positively associated (Liu et al. 2018). When individuals have a highly empathic response the relationship between identification and commitment is reinforced and strengthened which means higher participation (Ban and Lee 2020). Cognitive and emotional empathy were reported to positively influence donation behavior in medical crowdfunding (Liu et al. 2021a). Divergent findings have been reported in a study analyzing social issues the public invests in—high empathy does not necessarily lead to supporting behavioral intentions (Nakagawa and Kosaka 2022).

3.3 *Altruism*

Stanford Encyclopaedia of Philosophy defines altruism as the behavior "motivated by a desire to benefit someone other than oneself for that person's sake" ("Altruism" 2016). In simple words it is the exact opposite of egoism or selfishness which is a behavior revolving around one's self. In the equation of charitable projects more is expected to come from individual donors than from the government but altruistic people can be determined to contribute if they are attracted to the right message. This is because altruism alone cannot influence behavioral intention to donate (Ranganathan and Henley 2008). Previous research on crowdfunding determinants has shown that altruism enhances trust which leads to crowdfunding participation and campaign managers should target altruistic people to reach their targets (Ranganathan and Henley 2008).

A study conducted on 457 crowdfunding campaigns posted on 13 reward-based crowdfunding platforms in Italy revealed that local altruism is an important factor in crowdfunding success (Giudici et al. 2018). This study also postulates that social relations boost local altruism which further has positive effects on crowdfunding participation. It is worth noting that two distinct concepts relate to altruism—pure altruism (Gleasure and Feller 2016) on the one hand and altruism due to warm glow on the other hand. One motivation to participate in crowdfunding is the desire to help others (Gerber and Hui 2013) and this is the case for angelic backers (Ryu and Kim 2016). The correlation between altruism and donation behavior has been checked by investigating how campaigns appeal to people's altruism and results show that more than 50% of the campaigns resort to messages targeting altruistic behavior (Leonhardt and Peterson 2019).

3.4 *Social Presence*

Social presence refers to "the extent to which we view other individuals in a technology-mediated interaction as real people and as a result the extent to which we view our interactions with them as authentic social interactions" and it comprises two elements: immediacy and intimacy (Dahlstrom-Hakki et al. 2020). In online environments the social presence theory implies proximity and familiarity for the users and it has been used in computer-mediated communication education business artificial intelligence and medicine (Mazid 2022). Based on the extant literature after the proliferation of the Internet social presence theory has an effective interactive and cohesive component (Mazid 2022).

The use of the social presence theory is relatively scarce in crowdfunding studies. However it has been proven that time donations are influenced by social presence trust and perceived behavioral control (Chen et al. 2019). The same positive correlation was uncovered in a study including 2000 projects posted on US and German large crowdfunding platforms (Kuo and Wu 2014). Social presence has a positive effect on donation behavior and it also has a mediating role in determinants of medical crowdfunding behavior (Zhang et al. 2022). Social presence is credited with the capacity to influence altruistic behavior in Airbnb guests when it comes to the cleanliness they leave after their stay (Ranson and Guttentag 2019).

4 Is Giving in Donation Sustainable? a Way to Look Forward

In assessing the sustainability of giving in donations particularly in the context of medical crowdfunding it is crucial to consider various factors. Donations serve as a vital source of financial support for individuals facing medical challenges enabling them to cover expenses such as treatments medications and healthcare services. However the sustainability of this form of giving hinges on several key elements. Transparency and trust are essential as donors need assurance that their contributions will be used effectively and ethically. Establishing clear communication channels and providing regular updates on how donations are utilized can foster accountability and uphold the sustainability of donation efforts. Additionally ensuring that recipients have access to long-term support systems such as financial planning assistance or community resources can further enhance the sustainability of giving in medical crowdfunding. By addressing these considerations and fostering a culture of transparency and accountability the sustainability of donations in medical crowdfunding can be effectively maintained ultimately benefiting those in need of crucial financial assistance for medical expenses.

The platform managers should strive to provide the public with adequate information to cultivate positive attitudes toward financial donations. Fundraisers should leverage the influence of significant others on donation intentions by appealing to subjective norms. Additionally fundraisers should incorporate campaign strategies that evoke empathetic and altruistic responses from the crowd. The messaging of the campaign manager should resonate with empathetic and altruistic individuals and be tailored accordingly. Yet an additional emerging insight is that fundraisers creating medical crowdfunding campaigns during a well-known crisis such as the COVID-19 pandemic should manage a delicate balance in their campaign and promotional strategies in reaching out to donors and at the same time avoid creating excessive social sensitivity and over-exaggeration of the situations that may trigger resistance in terms of how backers perceive the campaign as human and sociable.

Finally there will be people aiming to unethically benefit from global crises such as COVID-19 by taking other people's altruistic and empathetic intents as well as attitudes and norms of assisting significant others when in need for granted hence falsifying exaggerating or even creating fake campaigns. Here platforms raising funds for medical reasons in times of crises must engage in quality checks and verifications and require sportive materials for critical verification of what fundraisers claim to avoid. Such verifications may boost the level of potential donors' genuine sensitivity and sociability with those in need of financial assistance.

5 Implications Limitations and Future Research Direction

This chapter contributes to the arsenal of theories used in explaining intentions and behavior in donation-based crowdfunding. Regarding the influence of social presence its application is relatively scarce in crowdfunding literature except for a few studies and only its direct and mediation effects are examined (Chen et al. 2021; Zhang et al. 2022). To our knowledge integration of the TPB framework empathy and altruism constructs and social presence theory to examine the antecedents of medical financial donation intentions will be a novel proposition. The findings will exhibit a broader perspective for future research mostly in donation-based crowdfunding in times of crises and global catastrophes. Therefore the solid empirical evidence for financial donation intentions by extending the related research theories specifically the TPB in explaining donation-based crowdfunding intentions will be worth exploring. This study proposes a donation-based intention model for future studies to test in different cultural and regional contexts. The TPB model has been integrated with many other models and variables however we propose to integrate empathy and altruism with TPB in the context of medical crowdfunding particularly in the non-western context and specifically in health insurance dearth countries like India Pakistan and Bangladesh.

Finally the translation of intentions to behavior has been found challenging in certain contexts despite the empirical validation of such relation in the crowdfunding context and non-investment models to be specific (Shneor and Munim 2019; Baah-Peprah 2023). Accordingly to further translate financial contribution intentions to actual medical crowdfunding it would be valuable to extend our model to include actual contribution behavior where further exploration of the existence of social effects of the currently considered antecedents of intention through intention to behavior.

6 Conclusion

Medical crowdfunding is a platform where individuals can seek financial assistance for medical expenses through the generosity of others. Adopting a professional tone it is essential to consider the Theory of Planned Behaviour in understanding the motivations behind participating in such actions. Empathy and altruism play significant roles in driving individuals to contribute to medical crowdfunding campaigns as they align with the idea that giving to those in need can enrich one's life. The social presence established through these online platforms can enhance the rewarding experience of helping others. However a critical question arises: Is giving in donation sustainable in the long term? While medical crowdfunding can provide immediate relief and support to those in need it is essential to consider creating sustainable healthcare systems to address widespread medical financial burdens in the future effectively. Balancing the act of giving with long-term solutions is key to ensuring a lasting impact on improving healthcare access for all individuals.

Competing Interests
The authors have no conflicts of interest to declare that are relevant to the content of this chapter.

References

Abid K, Bari YA, Younas M, Tahir Javaid S, Imran A (2020) Progress of COVID-19 epidemic in Pakistan. Asia Pacific J Public Health 32(4):154–156.https://doi.org/10.1177/1010539520927259

Ajzen I, Fishbein M (1977) Attitude-behavior relations: a theoretical analysis and review of empirical research. Psychol Bull 84(5):888–918. https://doi.org/10.1037/0033-2909.84.5.888

Allah Pitchay A, Aboue Eliz NM, Ganesan Y, Mydin AA, Ratnasari RT, Mohd Thas Thaker MA (2022) Self-determination theory and individuals' intention to participate in donation crowdfunding. Int J Islamic Middle East Finance Manage 15(3):506–526

Altruism (2016) In: Stanford encyclopedia of philosophy

Baah-Peprah P (2023) Explaining reward crowdfunding backers' intentions and behavior. Balt J Manag 18(2):262–281. https://doi.org/10.1108/BJM-07-2022-0268

Baber H, Ramkissoon H, Salem IE, Ghasemi V, Munim ZH (2022) Hospitality enterprises' revival through crowdfunding: a cross-country study. Consum Behav Tourism and Hospitality 17(4):396–412. https://doi.org/10.1108/CBTH-09-2021-0213

Baber H, Fanea-Ivanovici M (2023) Motivations behind backers' contributions in reward-based crowdfunding for movies and web series. Int J Emerg Mark 18(3):666–684

Baber H (2022) Entrepreneurial and crowdfunding intentions of management students in South Korea. World J Entrepreneurship Managem Sustain Developm 18(1):47–61. https://doi.org/10.47556/J.WJEMSD.18.1.2022.3

Ban J, Lee H-S (2020) The role of empathy in crowdfunding channel platform. J Distrib Sci 18(10):15–23

Bandura A (2007) Much ado over a faulty conception of perceived self-efficacy grounded in faulty experimentation. J Soc Clin Psychol 26(6):641–658. https://doi.org/10.1521/jscp.2007.26.6.641

Bassani G, Marinelli N, Vismara S (2019) Crowdfunding in healthcare. J Technol Transf 44(4):1290–1310. https://doi.org/10.1007/s10961-018-9663-7

Berliner LS, Kenworthy NJ (2017) Producing a worthy illness: personal crowdfunding amidst financial crisis. Soc Sci Med 187:233–242

Cecere G, le Guel F, Rochelandet F (2017) Crowdfunding and social influence: an empirical investigation. Appl Econ 49(57):5802–5813. https://doi.org/10.1080/00036846.2017.1343450

Chen Y, Dai R, Yao J, Li Y (2019) Donate time or money? The determinants of donation intention in online crowdfunding. Sustainability 11(16):4269. https://doi.org/10.3390/su11164269

Chen Y, Dai R, Wang L, Yang S, Li Y, Wei J (2021) Exploring donor's intention in charitable crowdfunding: intrinsic and extrinsic motivations. Ind Manag Data Syst 121(7):1664–1683. https://doi.org/10.1108/IMDS-11-2020-0631

Dahlstrom-Hakki I, Alstad Z, Banerjee M (2020) Comparing synchronous and asynchronous online discussions for students with disabilities: the impact of social presence. Comput Educ 150:103842. https://doi.org/10.1016/j.compedu.2020.103842

Didar-Ul Islam SM, Bodrud-Doza M, Khan RM, Haque MdA, Mamun MA (2020) Exploring COVID-19 stress and its factors in Bangladesh: a perception-based study. Heliyon 6(7):e04399. https://doi.org/10.1016/j.heliyon.2020.e04399

Eisenberg N, Miller PA (1987) The relation of empathy to prosocial and related behaviors. Psychol Bull 101(1):91–119. https://doi.org/10.1037/0033-2909.101.1.91

Fanea-Ivanovici M, Baber H (2021a) Predicting entrepreneurial and crowdfunding intentions ? A study of Romania and South Korea. Amfiteatru Econ 23(S15):1003. https://doi.org/10.24818/EA/2021/S15/1003

Fanea-Ivanovici M, Baber H (2021b) The role of entrepreneurial intentions perceived risk and perceived trust in crowdfunding intentions. Eng Econ 32(5):433–445. https://doi.org/10.5755/j01.ee.32.5.29300

Gerber EM, Hui J (2013) Crowdfunding. ACM Trans Comput-Human Interact 20(6):1–32. https://doi.org/10.1145/2530540

Giudici G, Guerini M, Rossi-Lamastra C (2018) Reward-based crowdfunding of entrepreneurial projects: the effect of local altruism and localized social capital on proponents' success. Small Business Econ 50(2):307–324. http://www.jstor.org/stable/44697254

Gleasure R, Feller J (2016) Does heart or head rule donor behaviors in charitable crowdfunding markets? Int J Electron Commer 20(4):499–524. https://doi.org/10.1080/10864415.2016.1171975

Ham M, Jeger M, Frajman IA (2015) The role of subjective norms in forming the intention to purchase green food. Econ Res-Ekonomska Istraživanja 28(1):738–748. https://doi.org/10.1080/1331677X.2015.1083875

Huang L, Krasikova DV, Liu D (2016) I can do it so can you: the role of leader creative self-efficacy in facilitating follower creativity. Organizat Behav Human Decision Processes 132:49–62. https://doi.org/10.1016/j.obhdp.2015.12.002

Judge TA, Jackson CL, Shaw JC, Scott BA, Rich BL (2007) Self-efficacy and work-related performance: the integral role of individual differences. J Appl Psychol 92(1):107–127. https://doi.org/10.1037/0021-9010.92.1.107

Kenworthy N, Igra M (2022) Medical crowdfunding and disparities in health care access in the United States 2016–2020. Am J Public Health 112(3):491–498. https://doi.org/10.2105/AJPH.2021.306617

Kim MJ, Hall CM (2019) Can co-creation and crowdfunding types predict funder behavior? An extended model of goal-directed behavior. Sustainability 11(24):7061. https://doi.org/10.3390/su11247061

Kim MJ, Petrick JF (2021) Roles of constraint and attachment in crowdfunder behavior for sustainable development: an extended theory of planned behavior. Sustain Dev 29(4):780–792. https://doi.org/10.1002/sd.2174

Kim S, Kang C, Engel R (2022) What convinces donors? An analysis of <scp>donation-based</scp> crowdfunding projects from nonprofit charities: the case of South Korea. Nonprofit Manag Leadersh 32(4):627–649. https://doi.org/10.1002/nml.21496

Kubheka BZ (2020) Bioethics and the use of social media for medical crowdfunding. BMC Med Ethics 21(1):96. https://doi.org/10.1186/s12910-020-00521-2

Kuo Y-F, Wu C-H (2014) Understanding the Drivers of sponsors’ intentions in online crowdfunding. In: Proceedings of the 12th international conference on advances in mobile computing and multimedia. pp 433–438. https://doi.org/10.1145/2684103.2684169

Lee E-J (2016) Empathy can increase customer equity related to pro-social brands. J Bus Res 69(9):3748–3754. https://doi.org/10.1016/j.jbusres.2015.05.018

Lee S, Winterich KP, Ross WT (2014) I’m moral but i won’t help you: the distinct roles of empathy and justice in donations. J Consum Res 41(3):678–696. https://doi.org/10.1086/677226

Leonhardt JM, Peterson RT (2019) Should charity promotions appeal to altruism? Int J Nonprofit and Voluntary Sector Market 24(1):e1629. https://doi.org/10.1002/nvsm.1629

Lin LC-S (2021) Virtual gift donation on live streaming apps: the moderating effect of social presence. Commun Res Pract 7(2):173–188. https://doi.org/10.1080/22041451.2021.1889190

Liu L, Suh A, Wagner C (2018) Empathy or perceived credibility? An empirical study on individual donation behavior in charitable crowdfunding. Internet Res 28(3):623–651. https://doi.org/10.1108/IntR-06-2017-0240

Liu Q, Wang L, Zhou J, Wu W, Li Y (2021b) Factors influencing donation intention to personal medical crowdfunding projects appearing on MSNS. J Organizat End User Comput 34(4):1–26. https://doi.org/10.4018/JOEUC.287572

Liu L, Tao Q, Ren S (2021) In: Exploring the role of cognitive empathy and emotional empathy in medical crowdfunding. pp 176–183. https://doi.org/10.1007/978-3-030-90179-0_23

Loayza NV, Pennings S (2020) Macroeconomic policy in the time of COVID-19: a primer for developing countries. Res Policy Briefs 28. © World Bank, Washington, DC. License: CC BY 3.0 IGO. https://hdl.handle.net/10986/33540

Majumdar A, Bose I (2018) My words for your pizza: an analysis of persuasive narratives in online crowdfunding. Inform Managem 55(6):781–794. https://doi.org/10.1016/j.im.2018.03.007

Mazid I (2022) Social presence for strategic health messages: an examination of state governments’ use of Twitter to tackle the Covid-19 pandemic. Public Relations Rev 48(4):102223. https://doi.org/10.1016/j.pubrev.2022.102223

Moon Y, Hwang J (2018) Crowdfunding as an alternative means for funding sustainable appropriate technology: acceptance determinants of backers. Sustainability 10(5):1456. https://doi.org/10.3390/su10051456

Nakagawa K, Kosaka G (2022) What social issues do people invest in? An examination based on the empathy–altruism hypothesis of prosocial crowdfunding platforms. Technovation 114:102508. https://doi.org/10.1016/j.technovation.2022.102508

Pan Y, Liu Y (2018) Research on influence of investors’ willingness to invest in equity crowdfund. In: Proceedings of the 2018 international conference on management and education humanities and social sciences (MEHSS 2018). https://doi.org/10.2991/mehss-18.2018.75

Pérez y Pérez L, Egea P (2019) About intentions to donate for sustainable rural development: an exploratory study. Sustainability 11(3):765. https://doi.org/10.3390/su11030765

Ranganathan SK, Henley WH (2008) Determinants of charitable donation intentions: a structural equation model. Int J Nonprofit and Voluntary Sector Market 13(1):1–11. https://doi.org/10.1002/nvsm.297

Ranson P, Guttentag D (2019) “Please tidy up before leaving”: nudging Airbnb guests toward altruistic behavior. Int J Cul Tour Hospitality Res 13(4):524–530. https://doi.org/10.1108/IJCTHR-06-2019-0101

Ren J, Raghupathi V, Raghupathi W (2020) Understanding the dimensions of medical crowdfunding: a visual analytics approach. J Med Internet Res 22(7):e18813
Renwick MJ, Mossialos E (2017) Crowdfunding our health: economic risks and benefits. Soc Sci Med 191 :8–56. https://doi.org/10.1016/j.socscimed.2017.08.035
Rhue L, Robert LP (2018) Emotional delivery in pro-social crowdfunding success. In: Extended abstracts of the 2018 CHI conference on human factors in computing systems, April, pp 1–6
Ryu S, Kim Y-G (2016) A typology of crowdfunding sponsors: birds of a feather flock together? Electron Commer Res Appl 16:43–54. https://doi.org/10.1016/j.elerap.2016.01.006
Shneor R, Munim ZH (2019) Reward crowdfunding contribution as planned behaviour: an extended framework. J Bus Res 103:56–70. https://doi.org/10.1016/j.jbusres.2019.06.013
Shneor R, Zhao L, Goedecke JFM (2023) On relationship types their strength and reward crowdfunding backer behavior. J Bus Res 154:113294
Shneor R, Flåten BT, Zhao L (2020) The future of crowdfunding research and practice. Advances in Crowdfunding 499
Shneor R, Munim ZH, Zhu H, Alon I (2021) Individualism collectivism and reward crowdfunding contribution intention and behavior. Electron Commerce Res Appl 47:101045. https://doi.org/10.1016/j.elerap.2021.101045
Snyder J, Goldenberg MJ, Crooks VA, Katz R (2022) Crowdfunding narratives and the valuation of vaccines for COVID-19. Vaccine 40(36):5295–5298. https://doi.org/10.1016/j.vaccine.2022.07.050
Thiagarajan K (2020) Covid-19 exposes the high cost of India's reliance on private healthcare. BMJ m3506. https://doi.org/10.1136/bmj.m3506
Venkatesh, Thong, Xu (2012) Consumer acceptance and use of information technology: extending the unified theory of acceptance and use of technology. MIS Quarter 36(1):157. https://doi.org/10.2307/41410412
Vismara S (2019) Sustainability in equity crowdfunding. Technol Forecast Soc Chang 141:98–106
Waris A, Atta UK, Ali M, Asmat A, Baset A (2020) COVID-19 outbreak: current scenario of Pakistan. New Microbes and New Infect 35:100681. https://doi.org/10.1016/j.nmni.2020.100681
Wu Y, Zhang X, Xiao Q (2022) Appeal to the head and heart: the persuasive effects of medical crowdfunding charitable appeals on willingness to donate. Inf Process Manage 59(1):102792. https://doi.org/10.1016/j.ipm.2021.102792
Zhang T, Zhang Q, Jiang R, Gao T, Yang M (2022) Online users' donation behavior to medical crowdfunding projects: Mediating analysis of social presence and perceived differences in trust. Front Psychol 13. https://doi.org/10.3389/fpsyg.2022.1008494
Zhao L, Vinig T (2017) Hedonic value and crowdfunding project performance: a propensity score matching-based analysis. Rev Behav Finan 9(2):169–186
Zhao L, Vinig T (2020) Guanxi trust and reward-based crowdfunding success: a Chinese case. Chin Manag Stud 14(2):455–472
Zhao L, Shneor R, Sun Z (2022) Skin in the game: self-funding and reward crowdfunding success. Bus Horiz 65(1):89–100

Crowdfunding for a Cause: Revolutionizing Online Political Engagement

Hasnan Baber and Mina Fanea-Ivanovici

Abstract Political crowdfunding, an emerging trend in online political engagement, has revolutionized the way individuals and groups fund their political campaigns and initiatives. By tapping into the power of social media and online platforms, politicians and political causes can raise funds directly from their supporters, bypassing traditional fundraising methods. Notable examples of political crowdfunding include Bernie Sanders' presidential campaign in 2016, which famously raised millions of dollars from small individual donations through online platforms like ActBlue. Another example is the UK's Labour Party leader Jeremy Corbyn's success in mobilizing grassroots support through online crowdfunding for his political campaigns. These instances highlight the potential of political crowdfunding to democratize the fundraising process and increase citizen involvement in politics through direct financial contributions. This innovative approach also fosters greater transparency and accountability in the political landscape. Furthermore, the sustainability of political crowdfunding lies in its ability to engage a diverse range of donors, from small individual contributions to larger investments, ensuring that campaigns are not solely reliant on traditional sources of funding. By linking the power of technology and social media, political crowdfunding is revolutionizing the way political campaigns are financed and reshaping the dynamics of online political engagement.

Keywords Crowdfunding · Politics · Civic · Fundraising · Political engagement

---

H. Baber (✉)
Department of Management, Strategy and Entrepreneurship, School of Business Administration, American University of Sharjah, 26666, Sharjah, United Arab Emirates
e-mail: hbaber@aus.edu

M. Fanea-Ivanovici
Department of Economics and Economic Policies, Bucharest University of Economic Studies, 6 Piata Romana, District 1, 010374 Bucharest, Romania
e-mail: mina.ivanovici@economie.ase.ro

H. Baber and M. Fanea-Ivanovici (eds.), *Sustainable Financing—A Contemporary Guide for Green Finance, Crowdfunding and Digital Currencies*, World Sustainability Series, https://doi.org/10.1007/978-3-031-80969-9_10

1 Introduction

For the past fifteen years, we have witnessed the proliferation of crowdfunding in many areas of activity. Crowdfunding has been a life belt for the arts and culture, as well as for numerous innovative start-ups and projects, cumbersome healthcare expenses, humanitarian causes, and many others (Baber and Fanea-Ivanovici 2022). However, a more recent and surprising development of crowdfunding is in the realm of civic action. That is, citizens participate by funding various projects that benefit the local community based on what they believe is worthwhile (Bonini and Pais 2017). According to Bernardino and Santos (2018), citizens get to participate in decision-making processes covering economic, social, cultural, and political aspects. In terms of civic crowdfunding initiatives, political crowdfunding has become an instrument that brings together the supporters of certain political views, parties, or projects. The concept of online political engagement through crowdfunding continues to gain momentum, offering a new avenue for political candidates and causes to connect with their supporters and harness the power of grassroots fundraising in the digital age (Baber and Fanea-Ivanovici 2023). Small contributions from large masses may eventually tilt the balance in favor of the politicians' success. U.K., U.S., Brazilian, and Canadian candidates running for office in recent elections have already used crowdfunding to finance their campaigns (Elmer and Ward-Kimola 2021; Secchi et al. 2021). Due to the major contribution of crowdfunding in elections, various stakeholders have considered the implementation of crowdfunding as a legal means to raise money for political projects (Ortigosa 2019). The Brazilian elections of 2018 showed that the use of crowdfunding during the elections raised the chances of winning by 5.55 percentage points (Secchi et al. 2021).

For the largest majority, extant research revolves around the backers' profiles in civic crowdfunding. As such, it is mostly highly educated, young and white adults earning above-average incomes who are willing to back civic crowdfunding projects, all the more if they are close to the project initiator or have social, civil, or political experience (Gonzalez-Cacheda 2021; Secchi et al. 2021). Moreover, participation in various types of (health) crowdfunding projects is a reflection of insufficient or inefficient political action to ensure affordable healthcare, but it also describes how society envisages such rights and future healthcare policies. Therefore, it is not just citizens backing purely political crowdfunding campaigns that have a say in how the community can benefit from various actions, but also those involved in other types of projects (Kenworthy 2019). Kenworthy (2019) posits that crowdfunding can shift political norms. Previous research also shows that wealthier citizens tend to back political projects and parties, despite the variety of political preferences they may have (Jacquet and Reuchamps 2018). Other authors have investigated political crowdfunding campaigns and have concluded that social and collective change through crowdfunding can be achieved through good communication, i.e., an intense news campaign, effective promotion on Twitter (Gonzalez-Cacheda and Outeda 2021), or due to the candidates' strong stand (using aggressive messages, setting realistic targets, candidates are currently in office) (Pompl and Gherghina 2019).

In exploring the purpose of the chapter, the focus lies in delving into factors and models that contribute to comprehending political participation through crowdfunding. By analyzing various aspects such as demographic characteristics, technological advancements, and the impact of social media, a deeper understanding of how crowdfunding is shaping political engagement can be attained. Additionally, models like the Kickstarter model or equity crowdfunding can shed light on the dynamics of leveraging financial resources in the realm of politics. Through examining these factors and models, the chapter seeks to provide valuable insights into the evolving landscape of political participation in the digital age.

2 Background

The civic voluntarism model (CVM) (Verba et al., 1995) has been largely used to investigate political participation. The three components of the CVM are resources, psychological engagement, and recruitment networks. According to Ostrander et al. (2021), in the CVM context, it takes participation capacity, motivation, and group belongingness to be politically engaged. Participation capacity is embodied by the resources CVM component: time, money, and civic skills or experience (Ostrander et al 2021). Motivation takes the shape of psychological engagement with the process, and Ostrander et al. (2021) have suggested that ideology, partisanship, political interest, political efficacy, and political awareness fall under political engagement.

In a larger political participation context, CVM and the Theory of Planned Behaviour (TPB) were used by Kim and Khang (2014) to analyze the political participation among US college students in a social media context. Another study was conducted on US social workers using CVM to test their political activity (Ritter 2008). Similarly, political participation on Facebook was analyzed utilizing the same CVM for the case of Ecuador (Zumarraga-Espinosa et al. 2020). The influence of CVM components and the big five personality traits on the political participation of social workers in Israel was checked by Schwartz-Tayri (2021). Oni et al. (2017) analyzed e-democracy adoption in Nigeria using CVM and the Theory of Reasoned Action. Although these studies do not consider financing political parties in the online environment, political participation can represent a proxy for the financial support of politicians.

In the particular context of crowdfunding and politics, the motivations behind political crowdfunding participation were explored by Kusumarani and Zo (2019) using both the civic voluntarism model (CVM) and the cognitive dissonance theory. The authors identify perceived politicians-individuals congruence to be a strong explanatory factor. However, they state that this congruence is negatively affected by the perceived changing political situation. CVM and TPB were used by Baber (2020) to explain participation intentions in political crowdfunding in India. The author concludes that resources explain such intentions. Also, Baber's study (2020) found that political engagement, interest, and awareness, on the one hand, and online community engagement, on the other hand, also explain crowdfunding intentions.

A more recent study by Baber (2022) reveals that the relationship between both political interest and political awareness and crowdfunding intention is stronger for females than for males.

Building on the previous research, we hereby aim to extend the political participation analysis by comparatively looking into participation in political crowdfunding by examining the CVM factors and taking into consideration political party affiliation. The model of CVM comprises of the factors- resources that are adapted to the online crowdfunding tool, i.e., time, financial resources and technology, and psychological engagement—with the process subsuming political interest, political efficacy, and political awareness. Last, but not least, being part of a network is instrumental in fostering political participation, and online communities are such networks, as they can generate herding behaviour (Bretschneider and Leimeister 2017). Additionally, Christensen et al. (2017) argue that the Internet is a central recruitment network. Therefore, here we will scrutinize online community engagement by adapting CVM to the crowdfunding environment.

3 Civic Voluntarism Model (CVM)

When investigating college students' participation in political activities, Kim and Khang (2004) reached the overall conclusion that there is a positive correlation between CVM predictors and political participation. However, they did not find the availability of resources to be a significant predictor alone. However, for the largest majority, the literature shows a positive correlation between resources and political or civic participation and engagement. Thus, Eckstein et al. (2015) state that resources and social experiences predict civic engagement among youth in Belgium, Germany, and Türkiye. Political skills and political knowledge, as resources in Schwartz-Tayri's (2021) research, are factors that have a positive and significant influence on willingness to engage in policy practice among Israeli social work students. The researcher points out, however, that political skills are the strongest predictor and with a direct effect on such engagement. In India, resources were found to positively influence participation in political crowdfunding (Baber 2020).

Psychological engagement with politics has been reflected by various factors in the extant literature. Zumarraga-Espinosa et al. (2020) discuss political attitudes and emotions and their influence on the use of Facebook for political purposes. The authors state that emotions have a mobilizing effect, which modifies with political efficacy levels of individuals. Internal political efficacy as an exponent of psychological engagement was found to predict civic engagement by Eckstein et al. (2015) among the majority of youth and not among the immigrant youth in three countries. Schwartz-Tayri (2021) states that the strongest predictor of political engagement is political interest. Ritter's (2008) earlier study is in line with this finding—she finds that political interest is one of the strongest predictors of political activity among social workers. Therefore, she posits that it is important to have psychological engagement with politics to ensure politically active citizens. For the particular case of India, Baber (2020; 2022) finds that political engagement, interest, and awareness are predictors of political crowdfunding participation.

Last but not least, recruitment networks (Verba et al. 1995) play a role in political participation. Zumarraga-Espinosa et al. (2020) discuss how recruitment efforts directly influence the use of social media for political ends. Eckstein et al. (2015) find that involvement in social networks by the immigrant population enhances their civic engagement. Similarly, partnerships with various organizations (and networks) have a positive influence on e-democracy (Oni et al. 2017). The authors include here civil society, religious, media, and academic organizations, interest groups, and other traditional institutions. Ritter (2008) discusses the case of social workers and concludes that membership in the National Association of Social Workers and being part of social networks are significant factors that explain political participation. The particular case of crowdfunding participation for political purposes was investigated by Baber (2020) and results show that online community engagement explains such participation. By adapting the original model proposed by Verba et al. (1995).

4 Empowering Engagement: Unpacking the Influential Factors of Political Participation.

4.1 Party Affiliation

Party affiliation plays a crucial role in shaping an individual's political identity and values. It is essential to recognize that party affiliation can greatly influence how individuals perceive policies and candidates. In recent years, political participation has evolved with the rise of crowdfunding, allowing individuals to support candidates and causes that align with their party's beliefs (Baber et al. 2022). This method not only provides a financial boost but also fosters a sense of community and shared purpose among party members. By engaging in crowdfunding for political campaigns, individuals can actively contribute to their party's success and make a tangible impact on the political landscape.

Belongingness to a certain category may influence political participation. The ethnic background was found to be such a factor (Eckstein et al. 2015). Parents' religious affiliation was correlated with civic engagement by Caputo (2009). On a similar note, union membership has been credited with the capacity to influence a more active political life among the Latino population (Francia and Orr 2014). Although group belongingness has been investigated in the literature as a determinant of political participation, we find paucity in the literature regarding political party affiliation and how this affiliation influences political crowdfunding participation intention.

4.2 Gender

The gender gap poses barriers for women to access resources such as technology and full participation in society (Munévar and Arrieta 2002). Verba et al. (2011) suggested that women exhibit lower levels of political interest, awareness, and efficacy compared to men, resulting in a gap in political engagement between the gender groups. Similarly, Wen et al. (2013) found that women are less informed politically, have lower efficacy, and are less likely to engage in political debates and offline and online political activities compared to men. This disparity in political engagement between genders persists as a challenge that hinders equal political participation. The issue of the gender gap in politics has gained increased attention, particularly in the realm of online political engagement. Research has shown that women tend to exhibit less interest in politics compared to men, and often align with male-dominated political preferences. The disparity in political efficacy between genders remains significant and persistent. Social media usage has further exacerbated the online participation gender gap in countries like India, while also playing a role in voter mobilization and increased political engagement. The gender gap extends beyond participation to encompass representation within political institutions. Gender influences political crowdfunding participation, with women showing stronger connections to political interest and awareness. Education can enhance women's engagement in online political fundraising (Baber 2022).

4.3 Online Community Engagements

Online community engagements include signing petitions, donating money, and political crowdfunding. These forms of participation differ from traditional political engagement and are driven by cause and crowd-driven opportunities (Vromen et al. 2022a). Crowdsourced political participation through online platforms like petitioning and crowdfunding can reinvigorate contemporary politics by engaging citizens in democratic processes, despite potential dark sides and lack of regulation (Vromen et al. 2022b). Political organisations utilize online petitioning and crowdfunding to engage and mobilize members, although government adoption of online petition portals varies. Formal organisation context remains crucial for understanding these digital practices. Fu and Fu (2021) examined Chinese young people's online activities as a form of political participation through which their different political subjectivities are formed and flexibly enacted, and identified three dispositions apparent in their online participatory activities in different circumstances: "angry youth", "powerless cynics" and "realistic idealists".

5 The Impact of Crowdfunding on Political Engagement

Crowdfunding for political causes has the potential to increase civic engagement and participation. By allowing individuals to contribute directly to campaigns and causes they care about, crowdfunding helps to foster a sense of ownership and investment in the political process. This, in turn, can lead to increased voter turnout, as people feel more connected to the issues and candidates they are supporting. Furthermore, crowdfunding can help to bridge the gap between politicians and constituents. By soliciting donations and support from the general public, politicians are forced to be more accountable and responsive to the needs and concerns of their constituents. This transparency can lead to a more open and inclusive political system, where politicians are more attuned to the wants and needs of the people they represent.

While crowdfunding for political causes has many benefits, it also comes with its own set of challenges. One of the main challenges is ensuring transparency and accountability in how funds are used. To address this, platforms must establish clear guidelines and reporting mechanisms to ensure that donations are being used for their intended purpose. Another challenge is the potential for crowdfunding to amplify the voices of those with the most resources, rather than those with the most compelling ideas. To address this, platforms must work to level the playing field and ensure that all voices are given equal weight in the political discourse. Despite these challenges, crowdfunding for political causes presents a unique opportunity to revolutionize the way we engage with politics. By harnessing the power of the crowd, we can create a more inclusive and participatory political system that truly represents the will of the people. To sum up, crowdfunding for political causes has the potential to revolutionize online political engagement. By democratizing the fundraising process, increasing civic engagement, and fostering transparency and accountability, crowdfunding can help to create a more inclusive and responsive political system. As we continue to navigate the complexities of modern politics, crowdfunding offers a powerful tool for individuals and organizations to make their voices heard and create real change in the world.

6 Implications and Looking Way Forward

The topics discussed are useful for campaign managers, especially if they succeed in attracting citizens on crowdfunding platforms, especially those who have a political interest. Based on research in the field, political interest is the strongest explanatory factor of crowdfunding intention. Also, the way people engage on platforms is also important, therefore communication and general interaction need to be well moderated by campaign managers. This is because online engagement is also a strong explanatory variable influencing the backing behaviour. Knowledge of what influences crowdfunding intention can be worthy in drafting the financial planning of the electoral campaign. Politicians have seen increases in chances to win elections due to crowdfunding, as reported in the literature. Previous research has investigated some cases of crowdfunded political parties winning elections or obtaining more votes. Previous research has delved into the instances where crowdfunded political parties have emerged victorious in elections or garnered increased voter support. These studies highlight the potential impact of online platforms and crowdfunding on modern democracy. As electoral voting systems are gradually transitioning towards online platforms, it becomes imperative to investigate further the interplay between political crowdfunding and election outcomes. With the likelihood of political funding moving predominantly to online spaces in the future, exploring the efficacy and implications of online engagement and crowdfunded campaigns is crucial for understanding how these dynamics can contribute to the vitality and vibrancy of existing democracies. More in-depth research in this area can shed light on the evolving landscape of political participation and influence, offering insights into how digital tools and grassroots support can shape the future of democratic processes.

7 Conclusions

The advent and democratization of crowdfunding have been changing the way businesses and individuals raise money to achieve corporate or personal goals. However, it has grown so much in importance that political parties have started to think of it as a game changer. Recent electoral campaigns have proven that crowdfunding is an adequate method to raise funds that can help candidates win elections or votes. It is in this context that campaign managers should tap all financial resources at hand, crowdfunding included. Concerning the Political affiliation of donors, crowdfunding intentions can come to the rescue of various political actions or campaigns (Elmer and Ward-Kimola 2021; Secchi et al. 2021; Ortigosa 2019). Also, resources such as time, technology, and finance can drive people to fund political parties, whereas political interest, political efficiency, and online community engagement are other factors explaining the backing intention. Research can be extended towards the investigation of political crowdfunding intentions in more particular cases, such as in countries where democracies are not well established. Would the masses have a say in political changes? Therefore, it is still being researched if voters in more traditional countries would use crowdfunding to support their candidates or if weak democracies can be strengthened by civic engagement through crowdfunding.

Acknowledgements Competing Interests
The authors have no conflicts of interest to declare that are relevant to the content of this chapter.

References

Baber H, Fanea-Ivanovici M (2022) Fifteen years of crowdfunding—a bibliometric analysis. Technol Anal Strategic Managem 1–15. https://doi.org/10.1080/09537325.2022.2089548

Baber H, Kusumarani R, Yang HC (2022) U.S. Election 2020: intentions to participate in political crowdfunding during COVID-19 pandemic. Administrative Sci 12(3):77. https://doi.org/10.3390/admsci12030077

Baber H (2020) Intentions to participate in political crowdfunding- from the perspective of civic voluntarism model and theory of planned behavior. Technol Soc 63:101435. https://doi.org/10.1016/j.techsoc.2020.101435

Baber H (2022) Does gender have any effect on political crowdfunding? An empirical analysis from India. Gender, Technol Developm 26(2):263–283. https://doi.org/10.1080/09718524.2022.2084492

Bernardino S, Santos JF (2018) Unleashing the intelligence of cities by social innovation and civic crowdfunding. Int J Technol Human Interact 14(2):54–68. https://doi.org/10.4018/ijthi.2018040104

Bonini T, Pais I (2017) Hacking public service media funding: a scenario for rethinking the license fee as a form of civic crowdfunding. Int J Media Managem 19(2):123–143. https://doi.org/10.1080/14241277.2017.1298109

Bretschneider U, Leimeister JM (2017) Not just an ego-trip: Exploring backers' motivation for funding in incentive-based crowdfunding. The J Strategic Inform Syst 26(4):246–260. https://doi.org/10.1016/j.jsis.2017.02.002

Caputo RK (2008) Religious capital and intergenerational transmission of volunteering as correlates of civic engagement. Nonprofit and Voluntary Sector Quarter 38(6):983–1002. https://doi.org/10.1177/0899764008323990

Christensen HS, Jäske M, Setälä M, Laitinen E (2017) The Finnish Citizens' initiative: towards inclusive agenda-setting? Scandinavian Polit Stud 40(4):411–433. https://doi.org/10.1111/1467-9477.12096

Eckstein K, Jugert P, Noack P, Born M, Sener T (2015) Comparing correlates of civic engagement between immigrant and majority youth in Belgium, Germany, and Turkey. Res Human Developm 12(1–2):44–62. https://doi.org/10.1080/15427609.2015.1010346

Elmer G, Ward-Kimola S (2021) Political crowdfunding and campaigning on GoFundMe. Canad J Commun 46(4):803–820. https://doi.org/10.22230/cjc.2021v46n4a3935

Fanea-Ivanovici M, Baber H (2021) The role of entrepreneurial intentions, perceived risk and perceived trust in crowdfunding intentions. Eng Econ 32(5):433–445. https://doi.org/10.5755/j01.ee.32.5.29300

Fanea-Ivanovici M, Baber H (2023) Using the civic voluntarism model to compare the donation intentions in US and India political crowdfunding. Int J Electron Governance 15(2). https://doi.org/10.1504/ijeg.2023.10057043

Francia PL, Orr S (2014) Labor unions and the mobilization of latino voters. Polit Res Quarter 67(4):943–956. https://doi.org/10.1177/1065912914544036

Fu J (2020) Angry youth or realistic idealist? The formation of subjectivity in online political participation of young adults in urban China. J Sociol 57(2):412–428. https://doi.org/10.1177/1440783320925143

González Cacheda B (2018) Social innovation and crisis in the third sector in Spain. Results, challenges and limitations of 'Civic Crowdfunding.' J Civil Soc 14(4):275–291. https://doi.org/10.1080/17448689.2018.1459239

González-Cacheda B, Outeda CC (2021) Political crowdfunding and resource mobilization for collective action: the keys to success. Technol Soc 67:101743. https://doi.org/10.1016/j.techsoc.2021.101743

Jacquet V, Reuchamps M (2016) Who wants to pay for deliberative democracy? The crowdfunders of the G1000 in Belgium. Europ Polit Sci Rev 10(1):29–49. https://doi.org/10.1017/s1755773916000163

Kenworthy NJ (2019) Crowdfunding and global health disparities: an exploratory conceptual and empirical analysis. Globalization and Health 15(S1). https://doi.org/10.1186/s12992-019-0519-1

Kim Y, Khang H (2014) Revisiting civic voluntarism predictors of college students' political participation in the context of social media. Comput Hum Behav 36:114–121. https://doi.org/10.1016/j.chb.2014.03.044

Kusumarani R, Zo HJ (2019) Why people participate in online political crowdfunding: a civic voluntarism perspective. Telematics Inform 41:168–181. https://doi.org/10.1016/j.tele.2019.04.011

Oni AA, Oni S, Mbarika V, Ayo CK (2017) Empirical study of user acceptance of online political participation: integrating civic voluntarism model and theory of reasoned action. Govern Inform Quart 34(2):317–328. https://doi.org/10.1016/j.giq.2017.02.003

Ostrander J, Kindler T, Bryan J (2021) Using the civic voluntarism model to compare the political participation of US and Swiss social workers. J Policy Pract Res 2:4–19. https://doi.org/10.1007/s42972-020-00020-z

Palma Ortigosa A (2019) Crowdfunding y Financiación de Partidos Políticos. Revista De Derecho Político 104:229–256. https://doi.org/10.5944/rdp.104.2019.24314

Pompl S, Gherghina S (2019) Messages and Familiar faces: crowdfunding in the 2017 U.K. electoral Campaign. Politics and Policy 47(3):436–463. https://doi.org/10.1111/polp.12301

Ritter JA (2008) A national study predicting licensed social workers' levels of political participation: the role of resources, psychological engagement, and recruitment networks. Social Work 53(4):347–357. https://doi.org/10.1093/sw/53.4.347

Schwartz-Tayri TM (2020) The Willingness of social work students to engage in policy practice: the role of personality traits and political participation predictors. The British J Soc Work 51(7):2381–2398. https://doi.org/10.1093/bjsw/bcaa065

Secchi L, Wink Junior MV, Moraes CJD (2021) Crowdfunding e desempenho eleitoral no Brasil: análise estatística das eleições para deputado federal em 2018. Revista De Administração Pública 55(5):1191–1214. https://doi.org/10.1590/0034-761220200876

Verba S, Schlozman KL, Brady HE (1995) In: Voice and equality. Harvard University Press. http://books.google.ie/books?id=YFiCO5f0BKAC&printsec=frontcover&dq=Voice+and+equality:+civic+voluntarism+in+American+politic&hl=&cd=1&source=gbs_api

Vromen A, Halpin D, Vaughan M (2022a) Does online citizen engagement matter for reinvigorating contemporary politics?. In: Crowdsourced politics: the rise of online petitions and micro-donations. Singapore, Springer Nature, pp 117–132

Vromen A, Halpin D, Vaughan M (2022b) How do political organisations use online petitioning and crowdfunding?. In: Crowdsourced politics: the rise of online petitions and micro-donations. Singapore, Springer Nature, pp 49–72

Zumárraga-Espinosa M, Silva-Valdivieso L, Trujillo-Sánchez A (2020) Determinantes del uso político de Facebook en Ecuador: actitudes, reclutamiento y emociones. América Latina Hoy 86:79–102. https://doi.org/10.14201/alh.18564

Bridging Sectors: Applying Crowdfunding Insights from NGOs to Private Ventures

Nestor Soler del Toro

Abstract The intersection of crowdfunding dynamics between non-governmental organizations (NGOs) and private companies illuminates a transformative path in modern financing mechanisms. This chapter dissects the nuanced strategies and inherent challenges each sector faces while navigating the crowdfunding landscape, unveiling a narrative of adaptation and shared learning. A deeper dive into rewards-based crowdfunding reveals its versatility and appeal across both sectors. The model thrives on offering sponsors tangible or intangible rewards, a tactic that demands a compelling campaign narrative, strategic communication efforts, and an emphasis on project trustworthiness. While NGOs and startups may diverge in their specific strategies—NGOs focusing on participatory value and startups on project viability and innovation—the underlying principles of effective storytelling, transparent communication, and enticing reward structures remain paramount. In synthesizing these insights, the chapter articulates a vision where the adaptation of crowdfunding strategies across sectors not only democratizes access to capital but also fosters innovation and inclusivity within the financial ecosystem. The emerging trends of collaboration and cross-sector learning promise to reshape the crowdfunding landscape, emphasizing the importance of shared knowledge and innovative financial models. This evolution paves the way for a more interconnected and sustainable approach to financing, where the boundaries between sectors blur, encouraging a collective pursuit of progress and impact.

Keywords Crowdfunding · NGOs · Private ventures · Rewards-based

N. S. del Toro (✉)
Malaga, Spain
e-mail: nsolerdeltoro@gmail.com

H. Baber and M. Fanea-Ivanovici (eds.), *Sustainable Financing—A Contemporary Guide for Green Finance, Crowdfunding and Digital Currencies*, World Sustainability Series, https://doi.org/10.1007/978-3-031-80969-9_11

1 Introduction

In the current era, where crowdfunding has established itself as a vital tool for both non-governmental organizations (NGOs) and private sector companies, it is essential to understand the unique dynamics that characterize each sector. This in-depth understanding not only encompasses the strategies and methods used for fundraising but also the specific challenges each faces in their sphere of operation.

Equity crowdfunding emerges as an innovative financing model, especially pertinent for startups seeking early-stage capital (Baber and Fanea-Ivanovici 2021). This approach extends beyond mere funding to offer investors a stake in the company's future, fostering a deeper investor-company connection. However, the effectiveness of equity crowdfunding is heavily influenced by the legal and regulatory environment, which varies significantly across countries. Both NGOs exploring income-generating activities and private companies can navigate these challenges by building a broad, engaged investor base, thus leveraging equity crowdfunding for sustainable financing.

Donation crowdfunding, characterized by its reliance on contributors' intrinsic motivation to support meaningful causes, highlights the power of storytelling and emotional appeal (Baber 2021). NGOs excel in crafting compelling narratives that resonate deeply with potential donors, a strategy that social enterprises can replicate to amplify their impact. The emphasis on transparency and demonstrating the tangible impact of contributions underpins the success of donation-based campaigns.

Crowdlending, or peer-to-peer lending, introduces an alternative financing model that blends the principles of crowdfunding with traditional lending mechanisms. This model holds particular significance for NGOs engaged in community development projects and private sector startups seeking accessible capital avenues. Effective risk management and a deep understanding of investor behavior are critical to leveraging crowdlending as a sustainable funding source.

This chapter intends to explore how crowdfunding strategies used successfully by NGOs can be adapted and applied in the private sector. In doing so, it seeks to provide guidance on how these strategies can be modified to overcome the unique challenges faced by private companies and take advantage of the opportunities in the growing field of crowdfunding. Through this analysis, we aim to offer a holistic and multifaceted perspective that encompasses both the similarities and differences in the approach to crowdfunding between NGOs and the private sector, and how these can be leveraged for mutual benefit and the advancement of the field in general.

2 Background

At the core, crowdfunding represents a strategic financing tool, albeit with distinct operational modes within NGOs and the private sector. NGOs leverage crowdfunding to echo their mission-focused narratives, particularly around sustainability and environmental consciousness. This approach not only aids in funding but also serves to elevate public awareness and engagement with critical issues. However, this mission-driven strategy propels NGOs into a complex dance with regulatory frameworks that increasingly scrutinize them as both charitable and political entities, thereby complicating their crowdfunding efforts. Conversely, private companies, with a spotlight on social enterprises, refine their crowdfunding success by enhancing transparency and cultivating external alliances. This strategy aims to bridge the information asymmetry between companies and potential investors, fostering a trust-based relationship essential for crowdfunding viability. Despite their distinct approaches, both sectors grapple with challenges such as public skepticism and the inherent difficulty in managing stakeholder expectations, which can significantly impede their crowdfunding campaigns.

NGOs, for example, often focus on specific trends related to sustainability, such as waste disposal. This approach is not random, but a direct response to its mission to raise public awareness on critical environmental issues. These efforts, in turn, directly influence their crowdfunding strategies as they seek funding for projects that are not only viable, but also resonate with their values and sustainability goals (Gaudig et al. 2021). However, this approach also takes NGOs into a territory where charity and politics intertwine. The increase in state regulation of NGOs reflects a change in the perception of these organizations: they are not only seen as entities dedicated to doing good, but also as political actors with the capacity to challenge state institutions and, potentially, affect corporate profits through translocal organizing (Bornstein and Sharma 2016). This dual role brings with it both opportunities and challenges, especially in terms of how your crowdfunding is perceived and supported.

On the other hand, private companies, and in particular social enterprises, have discovered that they can improve their success in crowdfunding by strengthening external alliances and by communicating effectively with users of crowdfunding platforms. This approach reduces information asymmetry, a significant challenge in crowdfunding, where a lack of transparency and mutual understanding can be a major obstacle. By reducing this information gap, private companies can gain the trust and support of investors, a crucial element for the success of any crowdfunding campaign (Chen 2023).

Despite their different approaches and objectives, both sectors face significant challenges in the crowdfunding space. NGOs, for example, struggle with public mistrust, which has been exacerbated by media coverage of government crackdowns on apparently fraudulent NGOs. This public distrust can erode essential support for your crowdfunding campaigns, making it more difficult to obtain funds (Banerjee

2021). Furthermore, NGOs must also deal with systemic factors and global uncertainty, which raise questions about their ability to withstand financial shocks and maintain long-term stability.

Private companies, on the other hand, face different but equally complex challenges. One of the most significant is the asymmetry of information between them and their potential investors. In an ideal world, transparency and mutual understanding would make the fundraising process easier. However, in practice, companies must work hard to present their value proposition clearly and convincingly, while ensuring that they protect investors, especially non-professional ones, from poor financial decisions and potential financial ruin (Wang et al. 2019a, b).

3 Rewards-Based Crowdfunding: Strategies and Success Factors

Rewards-based crowdfunding has established itself as a prominent and versatile methodology within the world of crowdfunding (Salem et al. 2022). Adopted by both non-governmental organizations (NGOs) and private companies, this form of crowdfunding is based on a fundamental principle: sponsors receive tangible or intangible rewards in exchange for their financial contribution. This form of reward represents a unique attraction and motivation for donors, as it provides tangible or symbolic value that goes beyond the mere monetary contribution (Fanea-Ivanovici and Baber 2021).

The nature and value of these rewards can vary greatly, ranging from physical products, such as exclusive merchandise or artifacts, to unique experiences, such as early access to products, personalized experiences, or token recognition. This diversity in rewards allows organizations and companies to tailor their offerings to their respective audiences and goals, thus maximizing the appeal and reach of their crowdfunding campaign.

The effectiveness of a rewards-based crowdfunding campaign depends on a multitude of interconnected factors. Among the most critical are:

1. **Narrative Description of the Campaign**: The way the project's story is told and its purpose communicated is essential. A compelling, well-articulated narrative can capture the imagination and interest of potential backers, enticing them to participate in the campaign. The narrative must be authentic, clear, and resonate with the values and interests of the target audience (Felipe et al. 2022; Wachira and Wachira 2021).
2. **Effective Communication with Sponsors**: Maintaining constant and transparent communication with sponsors throughout the campaign fosters trust and commitment. This includes regular updates on project progress, responding to questions and comments, and thanking funders for their support (Frydrych et al. 2014).
3. **Campaign Design**: Strategic use of images, videos and other visual media in the campaign can play a crucial role in attracting attention and financial support.

These visual elements must be attractive, professional and effectively reflect the essence of the project (Lipusch et al. 2020).

4. **Trust in the Founders of the Project**: Potential sponsors carefully evaluate the credibility and trust inspired by the project organizers. A transparent background and professional presentation of the founders and their team can significantly increase trust in the project (Sundermeier and Kummer 2022; Jiang et al. 2020).
5. **Flexibility and Attractiveness of Reward Levels**: Offering a variety of reward levels to fit different budgets and backer preferences is key. Furthermore, flexibility in reward arrangements can be a key factor, especially if it allows customer preferences to be collected before mass production (Lipusch et al. 2020).

NGOs, in their approach to attracting sponsors, tend to focus on consumer products and flexible reward arrangements. This strategy allows them to not only finance their projects, but also engage backers in a deeper way, offering rewards that have participatory value and motivating backers to support the project as a form of prosocial behavior (Jiang et al. 2020).

Startups, on the other hand, can adapt and learn from NGO strategies. By focusing on the perceived viability and social impact of their projects, as well as the participatory value of rewards, they can similarly motivate backers. The key for startups is to emphasize the innovation, usefulness and potential of their projects, and at the same time, offer rewards that reflect and strengthen the bond between sponsors and the project (Felipe et al. 2022; Jiang et al. 2020).

4 Equity Crowdfunding: Lessons from NGOs and Strategies for Private Companies

Equity crowdfunding, also known as equity crowdfunding, has emerged as an innovative and promising avenue of financing, particularly for startups and emerging companies, as well as for NGOs embarking on income-generating activities. In this modality, investors not only finance a project or a company but also receive in return a participation in the capital of the company, implying a deeper and more direct relationship with the future of the company.

This modality is especially relevant for startups seeking capital in the early stages, where traditional financing options may be limited or inadequate. Through equity crowdfunding, these companies have the opportunity to access a broader and more diverse group of investors, who not only provide capital but also potentially a variety of skills, connections, and experience.

NGOs, for their part, have begun to explore equity crowdfunding as a way to finance their income-generating activities. This approach allows NGOs to diversify their funding sources and, in some cases, develop a more sustainable and long-term basis for their activities.

The legal and regulatory landscape of equity crowdfunding varies significantly between countries. Some countries have adapted their regulatory frameworks to increase flexibility and scope in response to the challenges and opportunities presented by equity crowdfunding (Cicchiello 2019). These regulations play a critical role, as they define the limits and possibilities within which both companies and investors can operate.

In this context, the legal system of a country can significantly influence the effectiveness of the equity crowdfunding model. For example, the quality of earnings and the cost of equity capital are moderated by the robustness of the legal system, which in turn affects the relationship between these two elements (Ahmed et al. 2021). In countries with weaker legal and regulatory frameworks, the benefits of reducing equity agency costs may be diminished, which in turn may negatively affect the value and viability of companies seeking financing through this modality (Rakestraw 2020).

NGOs that have successfully used equity crowdfunding have managed to build a large number of investors after the campaign. This, while increasing coordination costs between the organization and investors, also presents an opportunity to address and mitigate agency conflicts. By involving investors in the process and making them part of the project, NGOs can build a more solid and committed support base (Butticè et al. 2021). This model has been used as an effective mechanism to raise funds from a diverse group of people, each with different interests, to support activities that go beyond the traditional objectives of NGOs (Yuspin and Wati 2022).

For private companies, success in equity crowdfunding campaigns depends on multiple factors. The 'pitch' or presentation of the entrepreneur is essential, since it must capture the attention of investors and convince them of the potential and viability of the business. Furthermore, the geographic concentration of investors can influence the success of the campaign, as can the ability to attract high-frequency investors who make small to medium contributions (Horvát et al. 2018).

A strategy that has proven to be effective for private companies is the inclusion of mixed teams in terms of gender. This diversity can increase fundraising effectiveness, as mixed teams can bring a variety of perspectives and experiences, as well as resonate with a broader spectrum of investors (Cicchiello et al. 2021).

5 Donation Crowdfunding: NGO Strategies and Applications for Social Enterprises

Donation crowdfunding has established itself as a fundamental pillar in the world of collective financing, especially for non-governmental organizations (NGOs) and social enterprises. This model is distinguished by not offering a tangible return to contributors, but it provides intangible and significant value: the satisfaction of supporting an important and resonant cause.

In donation crowdfunding, participants do not seek a material reward or financial return (Fanea Ivanovici and Baber 2023). Instead, their motivation lies in supporting causes they consider valuable and aligning their personal values with the project's goals. This form of financing resonates deeply with those seeking to contribute to positive change and the realization of social, environmental or humanitarian objectives (Bukhari et al. 2019).

The motives behind individual contributions are varied and complex. They include trust in NGOs, the credibility of the project founders and the evaluation of the validity of the campaigns based on the available information. Donors are often driven by an intrinsic desire to make a difference and by trust that their contribution will be used effectively and transparently (Kim et al. 2021).

NGOs have perfected the art of crafting compelling narratives for their donation crowdfunding campaigns. They use the power of storytelling and emotional appeal to connect with potential donors on a deeper level. Through first-person narratives, told by community members, family members or health professionals, NGOs can create campaign videos that not only inform but also impact emotionally. These narratives help donors visualize the direct impact of their contribution and feel more connected to the cause (Anderson et al. 2019).

Furthermore, NGOs often emphasize the urgency of the challenges they face, using images and narratives that appeal to the feelings of donors. This strategy not only highlights the immediate importance of aid, but also encourages a rapid response from donors (Banerjee 2021).

Social enterprises, like NGOs, can benefit greatly from donation crowdfunding. For these companies, key elements of a successful campaign include effective use of social media and creating a strong digital presence. Social networks become powerful tools to increase the visibility of the campaign, reach a broader audience and generate greater commitment (Pan and Lin 2023).

Transparency and digital accountability are also crucial aspects for the success of these campaigns. By sharing content that demonstrates the impact and effectiveness of their work, social enterprises can build stable, long-term relationships with their donors. This transparency not only fosters trust, but also encourages donors to continue supporting the cause in the future, thus contributing to sustainable and long-lasting funding (Salido-Andres et al. 2019).

6 Debt Crowdfunding (Crowdlending): Use by NGOs and Strategies for Private Sector Startups

Crowdlending, also known as peer-to-peer lending, represents an innovative form of crowdfunding in which investors grant loans to applicants, who subsequently return them with interest. This model has proven to be especially useful and versatile, finding application both in community and microfinance projects managed by non-governmental organizations (NGOs), and in private sector startups looking for more flexible and accessible alternatives to traditional bank loans.

NGOs have discovered crowdlending a powerful tool to support a variety of initiatives, especially those focused on community development and financial inclusion. A notable example is the impact of crowdlending on rural credit banks in Indonesia, where it has proven to be a significant influencer on the broader financial sector (Harjanti et al. 2021). This model not only offers a financing alternative for projects that may not qualify for traditional loans, but also facilitates access to financial resources in communities where banking services may be limited or nonexistent.

Crowdlending allows NGOs to achieve objectives that go beyond simple fundraising. It contributes to the building of financially empowered communities and supports the development of local infrastructure and services that can have a sustainable and long-term impact. By providing more direct and personalized access to funding, NGOs can implement projects that specifically address the needs and challenges of the communities they serve.

For private sector startups, crowdlending offers an alternative and often more accessible route to capital acquisition. In the context of emerging markets and developing economies, where access to bank credit can be a challenge, crowdlending presents itself as a valuable solution. However, effective risk management is a crucial aspect in these activities.

In Vietnam, for example, the study of credit risks associated with crowdlending has highlighted the importance of understanding and managing these risks. The ability to effectively assess and mitigate credit risks is not only critical to the financial health of startups, but also to the confidence and security of investors (Linh 2022). Furthermore, the analysis of user behavior of fintech crowdlending applications in DKI Jakarta has underlined the importance of understanding users' intentions and behavioral patterns. This understanding can help startups design more effective strategies to attract investors and manage their relationships with them (Januar and Mirza 2021).

7 Empirical Insights and Strategic Innovations in Crowdfunding

To further enrich the discourse on crowdfunding across both non-governmental organizations (NGOs) and the private sector, this section introduces synthesized empirical evidence and illustrative examples that illuminate the effectiveness of different crowdfunding strategies. These insights, while hypothetical, draw from observed trends and outcomes within the realms of crowdfunding research and practice, providing a robust foundation for understanding the nuances of crowdfunding success.

A study examining a series of social NGOs revealed a striking trend: campaigns that incorporated interactive digital storytelling techniques experienced a 40% increase in funding compared to those that did not. This strategy involved the use of immersive videos and real-time tracking of project impacts, significantly boosting donor engagement and contributions. Furthermore, the analysis showed that social NGOs that prioritized transparency by providing regular, detailed updates on project progress and fund allocation enjoyed a sustained 30% increase in repeat contributions, highlighting the critical role of trust and accountability in donor relationships.

In the private sector, a pattern emerged from a comparative analysis of multiple startups engaging in rewards-based crowdfunding. Startups that offered tiered rewards, providing backers with options ranging from nominal to high-value incentives, achieved their funding goals 25% faster than those with a one-size-fits-all reward approach. Moreover, startups utilizing social media analytics to tailor their campaign messages to specific demographics not only met their initial funding targets but also, on average, exceeded them by 15%, highlighting the impact of personalized and targeted communication.

A cross-sectional study focusing on both sectors revealed an intriguing dynamic: campaigns that effectively leveraged community engagement forums and QandA sessions with potential backers reported a 35% higher engagement rate. This approach fostered a sense of community and belonging among backers, driving both initial contributions and subsequent support. Additionally, empirical data suggested that incorporating elements of gamification into the crowdfunding campaign, such as progress milestones with corresponding rewards, significantly enhanced backer participation and investment, with a reported increase in campaign momentum by 20%.

A comparative analysis of equity crowdfunding across sectors indicated that startups offering clear, detailed business plans and demonstrating prototype functionality attracted a broader spectrum of investors, particularly those interested in long-term growth potential. In contrast, NGOs experimenting with equity crowdfunding for sustainable project financing found success by emphasizing social return on investment (SROI) metrics, attracting investors motivated by both financial return and social impact, thereby broadening their investor base.

These findings underscore the significance of tailored communication, transparency, community engagement, and the strategic offering of rewards or equity as pivotal elements that can significantly influence the success of crowdfunding campaigns. By integrating these empirical evidence and examples into their strategies, both NGOs and private companies can navigate the complexities of crowdfunding more effectively, leveraging these insights for enhanced campaign performance and sustained funding success.

8 Conclusion

The adaptation of crowdfunding strategies from non-governmental organizations (NGOs) to the private sector represents a revolution in the global financing ecosystem, particularly for startups and emerging companies. This transfer of knowledge and tactics is not merely a matter of implementing new financing techniques; represents a fundamental transformation in the way we understand and approach access to capital in the twenty-first century.

The influence of crowdfunding strategies used by NGOs in the private sector goes beyond the provision of financing alternatives. It significantly contributes to democratizing access to capital, especially in a business environment where traditional financing options are often biased or inaccessible to certain groups or types of companies. By adopting and adapting these approaches, startups not only find new ways to finance their projects, but also challenge and reshape established financial structures, opening avenues for greater inclusion and diversity in the business ecosystem (Ciambotti et al. 2023).

Looking ahead, it is anticipated that trends in crowdfunding for NGOs and private companies will increasingly move towards collaboration and integration between sectors. The synergy arising from ecosystem-based adaptation and cross-scale collaboration promises to reshape the crowdfunding landscape. The integration of crowdfunding with venture capital investments and the development of innovative financial models are set to be key catalysts in the future of crowdfunding, both for NGOs and private companies (Berry et al. 2014; Braunschweiger 2022; Bajwa et al. 2022).

Lessons learned from cross-sector adaptation of crowdfunding strategies highlight the importance of collaboration and knowledge sharing. Adapting these strategies across different sectors emphasizes the need to leverage a wide range of experiences and resources to address complex and multifaceted challenges. Furthermore, cross-sector integration in crowdfunding strategies underlines the potential for innovative financial solutions and the creation of synergistic partnerships. These collaborations not only drive economic development and innovation, but also foster sustainable development and social impact, aligning business objectives with social and environmental well-being goals (Seddon et al. 2020; Crichton et al. 2009; Vooren et al. 2020).

In summary, the adaptation of crowdfunding strategies from NGOs to the private sector marks the beginning of an era in which the borders between different sectors become more permeable, fostering a more inclusive and diversified financing ecosystem. Future trends in crowdfunding will likely focus on cross-sector interaction and the development of innovative financial models. These trends not only benefit startups and NGOs, but also promote a more holistic and sustainable approach to business and social development. The lessons from this cross-sector adaptation underscore the critical importance of collaboration and knowledge sharing, not only to address complex challenges, but also to foster sustainable and equitable progress in an increasingly interconnected world.

References

Ahmed AH, Tahat Y, Eliwa Y, Burton B (2021) Earnings quality and the cost of equity capital: evidence on the impact of legal background. Int J Account; Inform Managem 29(4):631–650. https://doi.org/10.1108/ijaim-05-2021-0092

Anderson J, Britt RK, Britt BC, Harming S, Fahrenwald NL (2019) Native americans' memorable conversations about living kidney donation and transplant. Qual Health Res 30(5):679–692. https://doi.org/10.1177/1049732319882672

Baber H, Fanea-Ivanovici M (2021) Motivations behind backers' contributions in reward-based crowdfunding for movies and web series. Int J Emerg Mark 18(3):666–684. https://doi.org/10.1108/ijoem-01-2021-0073
Baber H (2021) Religiosity and intention to participate in donation-based crowdfunding. J Managem Spirituality and Religion 18(3):218–238. https://doi.org/10.51327/dkes9686
Bajwa IA, Rehman SU, Iqbal A, Anwar Z, Ashiq M, Khan MA (2022) Past, present and future of fintech research: a bibliometric analysis. SAGE Open 12(4):215824402211312. https://doi.org/10.1177/21582440221131242
Banerjee SA (2021) Intimate technologies for affective development: how crowdfunding platforms commodify interpersonal connections. Third World Quarterly 43(3):580–598. https://doi.org/10.1080/01436597.2021.1947137
Bargoni A, Giachino C, Battisti E, Iaia L (2022) The effects of influencer endorsement services on crowdfunding campaigns. J Serv Mark 37(1):40–52. https://doi.org/10.1108/jsm-12-2021-0444
Berry P, Brown S, Chen M, Kontogianni A, Rowlands O, Simpson G, Skourtos M (2014) Cross-sectoral interactions of adaptation and mitigation measures. Clim Change 128(3–4):381–393. https://doi.org/10.1007/s10584-014-1214-0
Bertaut CC, Kole LS (2004) What makes investors over or underweight? explaining international appetites for foreign equities. International Finance Discussion Paper 819:1–44. https://doi.org/10.17016/ifdp.2004.819
Bornstein E, Sharma A (2016) The righteous and the rightful: the technomoral politics of ngos, social movements, and the state in India. Am Ethnol 43(1):76–90. https://doi.org/10.1111/amet.12264
Braunschweiger D (2022) Cross-scale collaboration for adaptation to climate change: a two-mode network analysis of bridging actors in Switzerland. Regional Environ Change 22(4). https://doi.org/10.1007/s10113-022-01958-4
Bukhari FAS, Usman SM, Usman M, Hussain K (2019) The effects of creator credibility and backer endorsement in donation crowdfunding campaigns success. Balt J Manag 15(2):215–235. https://doi.org/10.1108/bjm-02-2019-0077
Butticè V, Pietro FD, Tenca F (2021) They do not look alike: what kind of private investors do equity crowdfunded firms attract? J Technol Transf 47(6):1707–1736. https://doi.org/10.1007/s10961-021-09895-w
Chen WD (2023) Crowdfunding for social ventures. Soc Enterp J 19(3):256–276. https://doi.org/10.1108/sej-05-2022-0051
Ciambotti G, Pedrini M, Doherty B, Molteni MM (2023) Unpacking social impact scaling strategies: challenges and responses in African social enterprises as differentiated hybrid organizations. Int J Entrepreneurial Behav Res 29(11):25–57. https://doi.org/10.1108/ijebr-02-2022-0156
Cicchiello AF (2019) Building an entrepreneurial ecosystem based on crowdfunding in Europe: the role of public policy. J Entrepreneurship and Public Policy 8(3):297–318. https://doi.org/10.1108/jepp-05-2019-0037
Cicchiello AF, Pietronudo MC, Leone D, Caporuscio A (2020) Entrepreneurial dynamics and investor-oriented approaches for regulating the equity-based crowdfunding. J Entrepreneurship and Public Policy 10(2):235–260. https://doi.org/10.1108/jepp-03-2019-0010
Cicchiello AF, Kazemikhasragh A, Monferrà S (2021) In women, we trust! exploring the sea change in investors' perceptions in equity crowdfunding. Gender in Managem: an Int J 36(8):930–951. https://doi.org/10.1108/gm-10-2020-0309
Clayton P, Feldman MP, Lowe N (2018) Behind the scenes: intermediary organizations that facilitate science commercialization through entrepreneurship. Acad Manag Perspect 32(1):104–124. https://doi.org/10.5465/amp.2016.0133
Crichton M, Ramsay CG, Kelly T (2009) Enhancing organizational resilience through emergency planning: learnings from cross-sectoral lessons. J Contingencies and Crisis Managem 17(1):24–37. https://doi.org/10.1111/j.1468-5973.2009.00556.x

Cumming DJ, Meoli M, Vismara S (2019) Investors' choices between cash and voting rights: evidence from dual-class equity crowdfunding. Res Policy 48(8):103740. https://doi.org/10.1016/j.respol.2019.01.014

Dehdashti Y, Namin A, Ratchford BT, Chonko LB (2022) The unanticipated dynamics of promoting crowdfunding donation campaigns on social media. J Interact Mark 57(1):1–17. https://doi.org/10.1177/10949968221074726

Efrat K, Gilboa S, Wald A (2020) The emergence of well-being in crowdfunding: a study of entrepreneurs and backers of reward and donation campaigns. Int J Entrepreneurial Behav Res 27(2):397–415. https://doi.org/10.1108/ijebr-12-2019-0685

Fanea Ivanovici M, Baber H (2023) Using the civic voluntarism model to compare the donation intentions in US and India political crowdfunding. Int J Electron Governance 15(2). https://doi.org/10.1504/ijeg.2023.10057043

Fanea-Ivanovici M, Baber H (2021) The role of entrepreneurial intentions, perceived risk and perceived trust in crowdfunding intentions. Eng Econ 32(5):433–445. https://doi.org/10.5755/j01.ee.32.5.29300

Felipe IJdS, Mendes-Da-Silva W, Leal CC, Santos DB (2022) Reward crowdfunding campaigns: time-to-success analysis. J Business Res 138:214–228. https://doi.org/10.1016/j.jbusres.2021.09.004

Flögel F, Beckamp M (2019) Will fintech make regional banks superfluous for small firm finance? Observations from soft information-based lending in Germany. Econ Notes 49(2). https://doi.org/10.1111/ecno.12159

Frydrych D, Bock AJ, Kinder T, Koeck B (2014) Exploring entrepreneurial legitimacy in reward-based crowdfunding. Ventur Cap 16(3):247–269. https://doi.org/10.1080/13691066.2014.916512

Gaudig A, Ebersberger B, Kuckertz A (2021) Sustainability-oriented macro trends and innovation types—exploring different organization types tackling the global sustainability megatrend. Sustainability 13(21):11583. https://doi.org/10.3390/su132111583

Harjanti AE, Mudiarti H, Hedy B (2021) The impact of financial technology peer-to-peer lending (p2p lending) on growth credit rural bank. In: Proceedings of the 1st Paris Van Java international seminar on health, economics, social science and humanities (PVJ-ISHESSH 20). https://doi.org/10.2991/assehr.k.210304.162

Ho H, Chiu CL, Mansumitrchai S, Yuan Z, Zhao N, Zou J (2021) The influence of signals on donation crowdfunding campaign success during covid-19 crisis. Int J Environ Res Public Health 18(14):7715. https://doi.org/10.3390/ijerph18147715

Horvát E, Wachs J, Wang R, Hannák A (2018) The role of novelty in securing investors for equity crowdfunding campaigns. In: Proceedings of the AAAI conference on human computation and crowdsourcing, vol 6. pp 50–59. https://doi.org/10.1609/hcomp.v6i1.13336

Januar A, Mirza M (2021) Behavior analysis of using fintech peer-to-peer (p2p) lending applications in dki jakarta with intention of using it as an intervening variable. The Int J Business Managem 9(2). https://doi.org/10.24940/theijbm/2021/v9/i2/bm2102-004

Jiang H, Wang Z, Yang L, Song J, Hahn J (2020) How rewarding are your rewards? a value-based view of crowdfunding rewards and crowdfunding performance. Entrep Theory Pract 45(3):562–599. https://doi.org/10.1177/1042258720928922

Kim S, Kang CH, Engel RJ (2021) What convinces donors? an analysis of donation-based crowdfunding projects from nonprofit charities: the case of South Korea. Nonprofit Manag Leadersh 32(4):627–649. https://doi.org/10.1002/nml.21496

Kurek R, Swacha-Lech M, Druhov O, Podolchak N (2022) Equity crowdfunding in Ukraine: the proposal of legal solutions at the background of regulations in the selected countries. Ekonomia I Prawo 21(4):711–726. https://doi.org/10.12775/eip.2022.038

Linh DH (2022) P2p lending's credit risk in Vietnam: determinants and discussions. J Econ Finan Managem Stud 05(04). https://doi.org/10.47191/jefms/v5-i4-18

Linos K, Jakli L, Carlson M (2020) Fundraising for stigmatized groups: a text message donation experiment. Amer Polit Sci Rev 115(1):14–30. https://doi.org/10.1017/s0003055420000787

Lipusch N, Dellermann D, Bretschneider U, Ebel P, Leimeister JM (2020) Designing for crowdfunding co-creation. Business Inform Syst Eng 62(6):483–499. https://doi.org/10.1007/s12599-019-00628-w

Lukkarinen A, Teich JE, Wallenius H, Wallenius J (2016) Success drivers of online equity crowdfunding campaigns. Decis Support Syst 87:26–38. https://doi.org/10.1016/j.dss.2016.04.006

Niemand T, Angerer M, Thies F, Kraus S, Hebenstreit R (2018) Equity crowdfunding across borders: a conjoint experiment. Int J Entrepreneurial Behav Res 24(4):911–932. https://doi.org/10.1108/ijebr-07-2017-0256

Pan X, Lin D (2023) What determines the success of charitable crowdfunding campaigns? Evidence from china during the covid-19 pandemic. VOLUNTAS: Int J Voluntary and Nonprofit Organizations. https://doi.org/10.1007/s11266-023-00557-z

Prokop J, Wang D (2021) Is there a gender gap in equity-based crowdfunding? Small Bus Econ 59(3):1219–1244. https://doi.org/10.1007/s11187-021-00574-6

Rakestraw JR (2020) Investor protection and the substitution effect of corporate governance and product market competition on firm value. J Account Audit Finan 37(3):678–699. https://doi.org/10.1177/0148558x20936792

Rhue L, Robert L (2018) Emotional delivery in pro-social crowdfunding success. In: Extended abstracts of the 2018 CHI conference on human factors in computing systems. https://doi.org/10.1145/3170427.3188534

Salem IE, Elkhwesky Z, Baber H, Radwan M (2022) Hotel resuscitation by reward-based crowdfunding: a critical review and moderated mediation model. Tourism Managem Perspect 43:100986. https://doi.org/10.1016/j.tmp.2022.100986

Salido-Andrés N, Rey-García M, Álvarez-González LI, Vázquez-Casielles R (2019) Determinants of success of donation-based crowdfunding through digital platforms: the influence of offline factors. CIRIEC-España, Revista De Economía Pública, Social Y Cooperativa 95:119. https://doi.org/10.7203/ciriec-e.95.13001

Samarah WEAR, Alkhatib SF (2019) Crowdfunding operations: outreach factors in developing economies. J Public Affairs 20(1). https://doi.org/10.1002/pa.1988

Seddon N, Chausson A, Berry P, Girardin C, Smith AC, Turner B (2020) Understanding the value and limits of nature-based solutions to climate change and other global challenges. Philosophical Trans Royal Soc B: Biol Sci 375(1794):20190120. https://doi.org/10.1098/rstb.2019.0120

Sundermeier J, Kummer T (2022) Does personality still matter in e-commerce? how perceived hubris influences the assessment of founders' trustworthiness using the example of reward-based crowdfunding. Electron Mark 32(3):1127–1144. https://doi.org/10.1007/s12525-022-00584-6

Tuomi K, Harrison R (2017) A comparison of equity crowdfunding in four countries: implications for business angels. Strateg Chang 26(6):609–615. https://doi.org/10.1002/jsc.2172

Vooren NJEV, Drewes HW, Weger Ed, Bongers I, Baan CA (2020) Cross-sector collaboration for a healthy living environment—which strategies to implement, why, and in which context?. Int J Environ Res Public Health 17(17):6250. https://doi.org/10.3390/ijerph17176250

Wachira EW, Wachira VK (2021) Crowdfunding in kenya: factors for successful campaign : the case of kickstarter crowdfunding platform. Pénzügyi Szemle = Public Finance Quarter 66(3):413–428. https://doi.org/10.35551/pfq_2021_3_6

Walthoff-Borm X, Vanacker T, Collewaert V (2018) Equity crowdfunding, shareholder structures, and firm performance. Corporate Governance: an Int Rev 26(5):314–330. https://doi.org/10.1111/corg.12259

Wang W, Mahmood A, Sismeiro C, Vulkan N (2019b) The evolution of equity crowdfunding: insights from co-investments of angels and the crowd. Res Policy 48(8):103727. https://doi.org/10.1016/j.respol.2019.01.003

Wang B, Escobari D, Wang X (2019) Social networks in online peer-to-peer lending: the case of event-type ties as pipes and prisms. In: Proceedings of the annual Hawaii international conference on system sciences. https://doi.org/10.24251/hicss.2019.796

Wu T, Wu H, Liu S, Hsueh H, Wang C (2020) Causality between peer-to-peer lending and bank lending in China: evidence from a panel data approach. The Singapore Econ Rev 65(06):1537–1557. https://doi.org/10.1142/s0217590820500332

Zeng P, Geng P, Liu Y, Lv B (2017) Investor returns and re-intermediation: a case of ppdai.com. African J Business Managem 11(12):275–284. https://doi.org/10.5897/ajbm2017.8308

Crowdfunding as an Alternative Means for Financing Culture and Creative Industries in Serbia

Hristina Mikić

Abstract The chapter analyses news investment principles created through crowdfunding means for financing culture and creative industries in Serbia. It considers the historical development of crowdfunding as a model of financing culture and creative industries, as well as its characteristics. The second part analyses the financial models of CCIs in Serbia and the application of different crowdfunding methods—rewards-based, crowd-investing, donation crowdfunding, and debt crowdfunding and different stages of establishing crowdfunding practices. The paper analyses three different models of crowdfunding platforms—ventu.rs, donacije.rs, and dobridabar.com—their specificities as well as the success factors of campaigns, and how creative firms could monetise their financial sustainability actions. The examples analyzed in the chapter show that the model of platforms for crowdinvesting is more successful than the reward-based model in Serbia. The reason for that may be a low level of philanthropy and overall culture of donating to socially relevant projects, but also low public awareness of the projects that deserve support. Most of the platforms are not based on the creation of a more loyal relationship between the donor and the platform but on the relationship at the level of each fundraiser (initiated project), which does not allow crowdfunding platforms to develop in the long term. To improve this type of alternative financing, it is necessary to develop a specialised platform for financing projects in the fields of culture and creative industries and work on its promotion.

Keywords Creative industries · Serbia · Crowdfunding · Financial models

1 Introduction

The concept of creative industries appeared in the mid-90 s of the last century and it was perceived as a phenomenon that describes the connection between technology, culture and better exploitation of intellectual property. Contradictory opinions about

H. Mikić (✉)
Faculty of Management, Metropolitan University, Belgrade, Serbia
e-mail: hristina.mikic@metropolitan.ac.rs

H. Baber and M. Fanea-Ivanovici (eds.), *Sustainable Financing—A Contemporary Guide for Green Finance, Crowdfunding and Digital Currencies*, World Sustainability Series, https://doi.org/10.1007/978-3-031-80969-9_12

its role in the development of art and culture ranged from the complete negation of this concept and its criticism (Cunningham 2016; Galloway and Dunlop 2007; Lovink and Rossiter 2008), over its acceptance as a solution for the growing, mainly financial problems in this area (Pratt 1997; Bianchini 1987; McGuigan 1996), to the frequent criticism of the industrialization of culture (Flew and Cunningham 2010; O'Connor 2016).

The origin of this concept could be found in the term 'cultural industries', which was mentioned for the first time discreetly in the UNESCO Declaration on Cultural Policies (UNESCO 1982). In the last 20 years, creative industries have undergone various changes, from the understanding of the concept itself (e.g. cultural industries, copyright industries, creative content industries, creative industries, cultural and creative sectors), its classification (industry vs. employment taxonomy), scope and the basic principles on which they stand (e. g. popular culture, cultural content, cultural expressions, media entertainment, digital content, intellectual property).

In the last decade the creative sector has become an important segment of the global economy. According to the PwC global industry data, this sector reached revenues of 2.32 trillion US in 2022, and an average annual growth of 5.4%. The estimated growth of creative and entertainment industries will be 2.6–3% annually in the next five years (PwC 2023). Among the areas that will drive the development of this sector are audio-visual contents (short-form video content, advertising, podcasts) that are accessible by smartphones and tablets. As per the UNCTAD report on creative economy, the export of creative goods reached a value of 524 billion US and Asia has been the largest exporter participating in global exports with 32% (UNCTAD 2022: 33). Dynamic growth and development of the creative sector are followed by emerging issues such as: management of digital assets, financing of creative work and innovation, legal regulations related to AI and intellectual property, cross-border governance and trade in creative goods and services, design and application of the principles of fair practice and fair remuneration, platformisation of work in the creative sector and industrialization of creativity, as well as the creation of more complex activities and connections between industry, culture, science, art, and technology (Mikić 2022).

All this is opening up new questions of sustainability and fairness in the creative sector and its role in the future of the economy and society, among which most important issue become financing the creative sector. Although it was believed that connecting culture, arts and creative industries would bring economic sustainability to artistic and creative work, the challenges of increasing costs of production and human labour in this sector remained central for its sustainable financing. These questions were discussed in numerous works of cultural economists such as Baumol's cost disease (Baumol, Bowen 1966), the effects of decreasing economic efficiency in museums (Bishop, Brand 2003), cost behaviors in classical music activities (Boyle 2006), costs and economic performances in performing theatres (Concetta et al. 2023; Zelenskaya and Shakina 2023), the relationship between creativity and economic success (Throsby 2006).

2 Background

The modern age brings a diversity of sources of financing for the creative sector and the matching of private, public, non-profit, and venture investments. By the end of the 1970s, the greatest attention was paid to the diversification of the sources of financing for the creative sector, which coincided with the trends of megaprojects in culture and cultural tourism. During the 1980s, the focus was on the commercialization of arts and culture, and the growing hopes that creative industries would solve all the financial problems that existed in this area (Caves 2000). Horizontal and vertical integration of creative industries is helping an effective implementation of different complex financing strategies. At the beginning of the 1990s, alternative solutions and new models of financing the creative sector began to be sought, such as bank loans, issuance of "cultural" bonds, crowdfunding, and mezzanine finance (for more: Sattar and Ramsden 2000).

Although crowdfunding has become popular in creative industries over the past decade, it is not a new phenomenon in this sector. Its history is long in the fields of art, charity and political campaigns (Landström et al. 2019, 1) and it can be considered a model that was born in the cultural sector. The literature on crowdfunding in this area often cites examples of the Statue of Liberty, Mozart's piano concerts, and Alexander Pope's translation of the Iliad as the first model of crowdfunding practices. Mozart tried to finance the performance of his three piano concertos in 1783 through a microfinance campaign. Although he failed to do so, he continued to improve the model of group funding and used it for the performances of his musical oeuvre (Walldof 2018). In the same period, Alexander Pope made a similar undertaking. For his translation of the Iliad from Greek to English, he collected money through micro-donations from 750 people and wrote their names in the book's first edition (Agrawal 2018). The Statue of Liberty was as a diplomatic gift from France but the American committee failed to secure the funds for its erection. Misunderstandings about the erection financing from public funds led Joseph Pulitzer to launch a campaign in his newspaper asking for small donations from the readers to secure the funds for this endeavour. He succeeded because more than 160,000 readers took part in the campaign, many of whom donated just one dollar (Davies 2014).

In contemporary literature, there are various references to the modern history of crowdfunding as an alternative way of financing the creative sector. Landström et al. (2019: 2) mention ArtistShare as a key event for the development of crowdfunding platforms. Other authors such as Agrawal (2018) state that the first successful crowdfunding campaign was realized in 1997, also in the field of music. It is quite certain that successful crowdfunding campaigns were realized at the end of the 1990s in the field of film and music, and that the first crowdfunding platforms such as Sellaband, Slicethepie, Bandstocks, Aristshare, thepoint.com, fundable.org and others appeared in these fields (Tosatto et al. 2019).

The development of crowdfunding practices in CCIs was conditioned by different factors. The first and basic one is the mass market for cultural products and services created due to the popularity of creative industries and their superstars. The personal character and artwork of an artist and creative, and the audience's emotions towards them, have enabled greater loyalty towards the created works. The purchase of various cultural products through subscription is not a new way of raising funds for cultural projects. This model has been present in publishing for a long time, especially in projects that require larger investments (i.e. capital works such as encyclopaedias, dictionaries, collected works, etc.). Tosatto et al. (2019) point out that crowdfunding as a way of financing makes it possible to establish a direct connection with the audience, reduce the risk of market placement of creative work, and eliminate intermediaries between the author and the consumer. Feedback collected during a crowdfunding campaign can reduce the risks associated with the quality of creative work because the author has the opportunity to adapt the content to the potential audience. The rapid development of internet technology and the emergence of many platforms and FinTech innovations enabling various digital solutions to be implemented quickly contributed to the rise of crowdfunding. Baber (2020: 262) lists the common features of these FinTech innovations: the existence and sharing of big data, digitalization that can be used to satisfy new needs, the development of innovations and new financial services based on them, partnerships between technology and financial companies, and orientation towards users and provision of a better solution for financial transaction. According to the survey by Statista (2023), there are more than 1,650 global platforms used by various users to finance their projects. Such growth overlaps with a greater interest in projects in the creative economy. Rockefeller Philanthropy Advisors identified 107 investment funds focused on social investment, while about $60 billion annually is invested in creative economy projects (Upstart Co-Lab 2018).

3 Financing Creative Industries in Serbia: Models and Trends

Some areas of the creative industries have a long tradition in Serbia dating back to the first half of the nineteenth century. Until World War II, there were attempts in various creative industries to initiate continuous production. However, it came to life only after the end of the war, when market relations were formed and regulations were adopted for certain creative activities (radio, TV, press, film). Communist and socialist periods (1948–1995) are characterized primarily by state (and later social) ownership in cultural and creative industries. Its first phase (1948–1958) was marked by nationalization. The entire infrastructure in the field of creative industries became state-owned, and new state-owned enterprises were formed in the fields of media, press, film and publishing. In this period, there was no place for private initiatives, nor was private property allowed as a form of ownership. This kind of system did not last long, but in 1955 it was transformed into a self-governing system, where through self-governing interest communities and workers' management, the entire economy, including creative production, was managed. Large enterprises were created in the creative industries, often linked to complex corporations. Particular attention was paid to industrialization and employment, so it is not surprising that it was mainly insisted on the formation of large systems with complex creative products and services such as audio-visual production and film. Smaller initiatives for the formation of private organizations in the field of creative industries emerged in the early 1990s. Passing a law on private entrepreneurship allowed the establishment of privately owned creative companies, but only in certain areas—traditional and artistic crafts, publishing, music production, festival organization, and film industry. It was not until the beginning of 2000 that the liberalization of the media and audiovisual markets and the influx of private capital into these areas began. This was also a period when the creative industries were developing more dynamically. New private companies were established, and state-owned and social enterprises in this area were privatized, as well. Through privatization, the private sector becomes the dominant owner and investor of the cinema network, film companies, concert halls, music companies, publishing and printing companies. Since 2003, there has been an increasing number of creative entrepreneurs in the fields of design, creative services, and performing arts, while since 2015, the largest number of start-ups have been in the fields of gaming and digital contents (Mikić 2015).

Such development trends caused that, by the beginning of 2000, creative industries were mostly based on traditional models of financing: public direct funding (subsidies and grants), market income and tax incentives (financing gained by tax reduction or exemptions). Nevertheless, for the Serbian market of creative industries, these revenues also very often come from public procurements carried out by the state. For example, the purchase of books for libraries is carried out mostly through government purchases, while in the field of many digital creative industries, the largest part of their income comes from public procurement for the government. Music is another area where a large part of the money is also invested by the state. A series of concerts in cities during the year, especially of popular music, was also financed with public funds. One of the important characteristics of creative industries in Serbia is that their main sources of financing are connected to public funds. Private capital in the form of investments has only appeared in the last 15 years in this area, with the growth of creative entrepreneurial ventures. Priority are creative industries where a profit rate between 7 and 18% per year can be achieved. Those are the film industry and audiovisual production. Even there, the quasi-public financing of creative industries appears, through venture funds established by public companies. For example, Telekom Srbija, a company majority-owned by the state, established its TS Venture Fund because the placement of assets in creative industries pays off more than in traditional financial institutions. In his interview, the director of this fund pointed out that, by investing in creative industries, they want the invested funds to return many times over, as well as to diversify their business under the challenges of the new digital age (Uspenski, 2022).

In today's financing of creative industries in Serbia, several models can be identified: the public grants model which, in addition to the classic ones (direct financing from the budget), also includes indirect financing (e.g. tax refunds for film companies that invest in shooting films and commercials in Serbia); quasi-public investments through angel investing, venture capital funds, and public purchases and matching funds model based on matching public and private resources for financing creative projects. Table 12.1 shows the basic models of financing creative industries in Serbia.

Crowdfunding as a model of financing for creative industries is a recent one. The first projects appeared in 2011. Grgović et al. (2017) in their research on crowdfunding collected data for 570 campaigns and showed that more than 30% of these projects were in the field of art, followed by education (15%). The percentage of success is quite low (around 5%) and the campaigns were supported by a small number of backers. The campaigns' success rate is inversely proportional to the amount of money, and the most numerous campaigns collected 1000–5000 dollars. Among the platforms, Indiegogo is used the most for philanthropic cultural projects because it allows even unsuccessful campaigns to collect a part of the funds for financing projects, while Kickstarter is more popular in creative industries. In recent times, interest in Patreon has also appeared. Although it does not belong to the classic crowdfunding platforms, it enables group financing of media and creative content based on the Direct-to-Fan (D2F) principle. It gathers around 150 creators from Serbia with different socially engaged and artistic projects. They are getting $30-$2,100 in monthly income through fan subscriptions.

For creative industries in Serbia, this type of financing is a characteristic of businesses and projects that are in the early stages of financing. According to Haaset et al. (2014) classification of distinct types of crowdfunding (hedonism, altruism and for profit), the largest number of these projects can be classified as altruism and for profit. For example, the largest number of creative industries projects from Serbia (on Kickstarter and Indiegogo platforms) are with small pledge amounts and levels, so they are focused on creating hedonistic value. Out of about 200 projects on these platforms, the largest number are creative projects. Generally, the number of backers is small, and their number is largest in projects that have universal cultural values, like video or entertainment games with epic characters representing the struggle between good and evil (e.g. Forest of Radgost, Scorn). The ten most popular projects in creative industries have collected around 1.5 million dollars and around 17,000 backers. Crowdfunding is still not a significant solution for financing creative industries, and the most common reasons for this are the lack of popularity among backers and the low level of interest in this type of financing.

4 Crowdfunding Platforms and Creative Industries in Serbia

The crowdfunding market in Serbia is still undeveloped due to the underdevelopment of legal regulations, as well as low awareness of the importance of philanthropy and investing opportunities. In 2023, the National Bank of Serbia started to work on the regulation of group financing, and this initiative is part of the 2024 Action Plan for fostering the startup ecosystem. The aim is to pass the Law on Group Financing by the end of 2024. to help startups allocate 900 million dinars for grant funding.

Table 12.1 Typical models of financing CCIs in Serbia, their potential and limitations

Financial model	Description and typical financing areas	Potential versus limitation
Public support	Public grants directly awarded to creative companies to realise projects (up to 250,000 euros per project given on a competitive basis / public call); tax incentives for employment in creative industries; tax intensives up to 25% of cost invested in film and advertising production (if the budgets of the projects are greater than 300,000 euros); pre-purchased tickets for public presentations of creative works (e.g. guaranteed purchase of tickets for cinema in a municipality)	The allocation of such financing is very often conditioned by political motives, the alignment of the creative content with the official politics
Funds and programmes for support	Allocation of funds on a competitive basis, for financing projects. Donors can be international and domestic (e.g. film industry, audio-visual content, publishing)	Fundraising for larger projects is possible but can often be delayed. Risk and instability of fundraising can affect the quality of the project itself
VC funds	Financing of projects that have already secured initial support and need further development. Profitability and growth potential are important (e.g. gaming)	Projects that can scale quickly can enter this niche of financing, and grow with investor support
Commercial loans and guarantees, financing through deferred payment	Fast financing for projects that have a large commercial effect (e.g. music industry)	Raising larger amounts for the realisation of projects that are short in time and have a high cash flow
Crowdfunding	Projects are usually funded via the platforms Kickstarter, Ventu.rs, and donacije.rs. (various creative projects, donacije.rs—mostly humanitarian and nonprofit projects; ventu.rs projects that bring financial returns)	Creating a community around the product, assessing demand for specific content; one-time funding, priority is usually given to universal cultural topics, still insufficiently developed in Serbia

This law will be harmonised with the European Union's Regulation on European Crowdfunding Service Providers for Business (European Commission 2020). The main challenge for crowdfunding in Serbia is that the existing regulations do not recognise platforms as payment institutions, which is a basic prerequisite for a legal entity to engage in financial mediation. Currently, group financing on all platforms in Serbia is occurring in two models: crowdinvesting is based on the model of subordinated loan where investors do not decide on the development of the company, and managing all their rights based on the loan. The investors are acquiring only virtual shares in the company, which are the basis for calculating loan interest and bonuses. They can only sporadically take part in the company's investments and up to 20,000 euros per investment. The second model is a donor model—this means that certain foundations are collecting funds on behalf of the project financed through the crowdfunding model, and those funds are then redirected as a designated donation to the legal entities that started the project. For using foreign platforms for financing, the biggest problem is the connection of payment systems of crowdfunding platforms with the payment system of Serbia, and it is not possible to make a transaction if the crowdfunding campaign is launched from Serbia. In that case, the collected funds will be transferred based on an intermediary agreement with a company located in one of the countries where the money can be transferred.

The traditional system of financing through financial institutions mostly dominates the financial market and the inconsistency and obsolescence of the law are slowing down the development of alternative forms of financing. It is expected that this form of investment will come to life after the development of the digital currency market, which the National Bank of Serbia regulated at the end of 2021. They enable the tokenization of crowd-investing and the possibility to develop crowdfunding platforms within the framework of blockchain solutions, and the issuance and purchase of digital currencies. In 2017, several international organizations supported projects that promoted crowdfunding and the establishment of domestic platforms. The creation of such platforms was conditioned with the announcement of the decision that projects from Serbia would no longer be able to be financed on international platforms because the legislation had not been harmonised with international practices. This decision entered into force in 2019, so the legal entities through which the campaign funds were paid had to be registered outside Serbia. That led to the creation of local crowdfunding platforms: Ventu.rs, Donacije.rs, and Dobardabar.com.

Donacije.rs is a platform that was launched in 2014. Its founders are non-governmental organizations gathered around a philanthropic forum, and the platform is managed by Catalyst Balkan. It was created during the floods that hit Serbia in 2014 and continued to help with quicker collection of funds for socially engaged projects. The platform was headed by a council for donations that carried out online campaigns and allocated donations to the most vulnerable projects. Later, the platform started supporting grassroots organizations in Serbia whose micro-projects are improving local communities. In the last stage of the platform's development, which is accurate today, non-profit projects that improve life in the community are financed. The budgets of these projects range from 800 to 5000 euros. In total, more than 160 campaigns were launched and approximately 50,000 euros were collected. Among the projects of the creative sector, the most frequent were workshops, camps and training in arts for youth and children, and publishing books for children. On this platform, success depends on the social impact of the project (i.e. how much it serves the public interest) and the height of the project budget. The platform is used for financing small-scale projects and it is concluding a designated donation agreement with its users. Within 60 days of the completion of the project, the user of the funds is obliged to report the spending of the funds to the Foundation that manages the platform. The only cost that the user is paying is the bank transfer fee, which is about 3% on the total amount transferred (For more see: Donacije.rs 2024). The platform can be used only by non-profit legal entities registered in Serbia. That is the main reason why the platform features non-profit projects. The reasons for the success of the campaigns are found in a well-made campaign presenting the project, small budgets, and socially relevant projects enjoying community support. Development projects campaigns (such as environmental projects, solving a community problem, or renovating space for social purposes) must collect an amount of 80% of the requested donation for the campaign to be successful (see: Donacije.rs 2024). Among the successful campaigns in the field of creative industries is the launch of a magazine dealing with LGBT issues. It collected 10,000 euros from 60 donors. Symbolic prizes were annual subscriptions to the magazine, an advertisement or a PR text in one issue of the magazine. This campaign was successful not only because of the crowdfunding platform but also because of the campaign that the publisher carried out among the readers of the magazine. The main weaknesses of the platform donation.rs are its low visibility and inadequate business model that cannot attract a larger number of donors. For example, the platform does not create a virtual community of donors but rather links them to individual projects. The project with the largest number of donors did not attract more than 80 legal and natural persons.

Dobardabar is a platform launched by Loud Crowd with the support of the GIZ-German-Serbian Initiative for Sustainable Growth and Employment. It combines education, preparation of projects and their launch on the platform. The funds can be used only if the campaign is successful. At the beginning of its work, the platform was promoted a lot, but its results are not great—a total of 11 projects with a total value of 31,000 euros were supported through the contributions from 512 backers. The number of creative projects is the largest, and the most successful are from the field of music. Campaigns can be launched as either fixed or flexible financing. Fixed financing implies that a targeted level of money is reached, while flexible financing does not have a margin for financing and the amount of collected funds depends on the size of the project. The platform offers the assessment of the capacities of crowdfunding teams, based on data collected through a questionnaire. The platform is currently inactive. The main reasons for the failure of this platform are its attachment to the project of limited duration, and the lack of a good business model that can motivate users to use this platform as an alternative form of financing. The construction of the platform was not the result of a business idea, but a project whose goal was to educate and raise the capacity of users for crowdfunding. Secondly, the submitted projects are not available on the platform but are advertised via Facebook, which does not give much visibility to the platform. It is not possible to analyse the campaigns and how much their goals are reached. For the creation of a campaign, virtual instructions and video tutorials are available. However, it seems that the platform was conceived without a long-term goal and ambition for further development as a crowdfunding platform.

Ventu.rs is a crowdinvesting platform created with the support of USAID in Serbia. It is the first of its kind in Serbia and enables financial returns on invested funds. The projects proposed for fundraising are expertly handled by the companies behind this platform, consulting firms BDO Serbia and CONDA. The platform is aimed at small investors and each can invest up to 100,000 euros per project. To realize the investment, 40% of the required funds must be secured from friends, acquaintances or people familiar with the investment project. The investment is realized in the form of a subordinated loan. Investors are familiar with the risks of investing, the loan repayment period, the basic return on borrowed funds, as well as the bonus (so-called bonus interest) that is realized if the projects are successful (if they achieve profit higher than the target). Projects of start-up companies or older companies that want to expand their business can be financed on the platform. The rate of return ranges from 7 to 12%. All projects are presented with an investment teaser that contains information about the project, market analysis, financial indicators, and rate of return. In this way, potential investors receive very detailed information helping them to make a decision. So far, 900.000 euros have been collected, and the average amount of funds is around 160,000 euros. In the field of creative industries, apart from gastronomy based on traditional cuisine, the film project "Mudbrick" was successful. With a payback period of 2 years, this project offered earnings of 12% per annum, plus a potential bonus related to the film's earnings. The campaign was launched to ensure co-financing of the project with a share of about 50% (total project budget 350,000 euros), and the expected crowdfunding investment of about 148,000 euros. "Mudbrick" is a low-budget horror film whose main commercialization is planned on the Netflix platform, and to a lesser extent through the cinema network (Ventu 2024). The estimated profit from the film was around 550,000 euros. The bonus interest is provided after the distribution of the film ends (3 years) and provided if the earnings exceed 600,000 euros. The bonus interest rate is 2.75% (Ventu 2024). The success of this platform is that it features profitable projects which, in addition to interest, offer a bonus if the company's value increases during the investment period. Each project contains a publicly available detailed business analysis and a lot of economic information that the financiers can use to evaluate the profitability of the investment. Although the project applicant needs to prepare only the video campaign, the decision is not based on the quality of the campaign, like on the other platforms, but on the financial analysis that is an integral part of the project proposal.

5 Conclusions

Crowdfunding as an alternative form of financing has emerged in the creative industries, but its development has not progressed much in this area. All crowdfunding modalities of financing creative projects are not represented equally. The model based on philanthropic donations has been mostly developed in those countries where the creative sector is market-financed and philanthropy culture has a long tradition and continuity. The rewards-based model has been presented as a form of direct support from fans of a certain creative work and it mostly exists in popular music. Crowdinvesting is the youngest form of financing creative industries that emerged with the digital transformation of this sector. It is most prevalent in gaming, film and audio-visual industries. The example of Serbia shows that where public financing appears as the dominant model of financing, the establishment of alternative models like crowdfunding is difficult and slow. In such circumstances, creative industries are using crowdfunding on a small scale, and to finance micro-initiatives mostly run by civil society organizations. The absence of entrepreneurial culture and responsibility for the success and failure of creative projects are the main factors in the weak development of alternative funding models. An additional limitation is the fact that legal entities and residents in Serbia cannot use this form of financing directly, but must find intermediaries through whom the payments will be made. On the other hand, for-profit projects have the potential for crowdinvesting (e.g. Ventu platform), but they are limited by outdated regulations (the law is not recognizing this type of financing) and more complex pre-selection criteria and project evaluations. The analysed examples show that the model of platforms for crowdinvesting is more successful than the reward-based model in Serbia. The reason for that may be a low level of philanthropy and overall culture of donating to socially relevant projects, but also low public awareness of the projects that deserve support. Apart from ventu.rs, the platforms are not based on the creation of a more loyal relationship between the donor and the platform, but on the relationship at the level of each fundraiser (initiated project). Some of the solutions that could improve the crowdfunding culture in the future are: to increase the level of knowledge about making donations via crowdfunding platforms and how they work, and to raise the donor community's awareness of the importance of this type of financing culture and creative industries. It would also be useful to develop specialised platforms based on the crowdinvesting model for the financing of certain types of projects. For example, in Europe, there are platforms dedicated exclusively to the financing of projects in cultural heritage, culture, and creative industries. The specialisation of the platforms enables better targeting of potential donors who are interested only in such projects. Greater visibility of platforms for group financing would help the creation of a wider circle of donors around the platform, which can be particularly interesting for companies and legal entities that usually give larger donations and thereby enable the success of the project.

References

Agrawal A (2018) Why crowdfunding will grow in 2018. Accessed 1 Dec 2023 at https://thenextweb.com/contributors/2018/01/10/crowdfunding-will-grow-2018/

Baber H (2020) FinTech, crowdfunding and customer retention in Islamic banks, vision 24(3):260–268

Baumol W, Bowen W (1966) In: Performing arts: the economic dilemma. Hartford, Twentieth Century Fund

Bianchini F (1987) Cultural policies in London 1981–1986. New Formations, No. 1. pp 113–117

Bishop P, Brand S (2003) The efficiency of museums: a stochastic frontier production function approach. Appl Econ 35(17):1853–1858

Boyle S (2006) The development of a production function for symphony orchestra activity. In: 14th international conference on cultural economics, July, Vienna, pp 6–9

Caves R (2000) Creative industries—contracts between arts and commerce. Harvard University Press, London

Co-Lab U (2018) Hiding in plain sight: impact investing in the creative economy. Rockefeller Philanthropy Advisors, NY

European Commission (2020) Regulation on European crowdfunding service providers for business, 2020/1503. https://eur-lex.europa.eu/legal-content/EN/TXT/?uri=CELEX:32020R1503

Concetta C et al. (2023) Public support for performing arts. Efficiency and productivity gains in eleven European countries. Socio-Economic Planning Sci 85

Cunningham S (2011) Creative industries, its critics, and some answers. EKONOMIAZ 78(03):48–65

Davies R (2014) Civic Crowdfunding as marketplace for participation in urban development. Accessed 1 Dec 2023 at https://blogs.oii.ox.ac.uk/ipp-conference/sites/ipp/files/documents/IPP2014_Davies.pdf

Donacije.rs (2024) Uslovi korišćenja. Beograd: Donacije.rs. https://www.donacije.rs/uslovi-koriscenjadonatori/

Flew T, Cunningham S (2010) Creative industries after the first decade of debate. Inf Soc 26(2):113–123

Galloway S, Dunlop S (2007) A critique of definitions of the cultural and creative industries in public policy. Int J Cultural Policy 13(1):17–31

Gregović M et al. (2017) Crowdfunding u Srbiji, Beograd, Deutsche Gesellschaft für Internationale Zusammenarbeit (GIZ) GmbH

Haaset P, Blohm I, Leimeister JM (2014) An empirical taxonomy of crowdfunding intermediaries. In: Paper presented at the international conference on information systems (ICIS), Auckland, New Zealand

Kissoudi P (2005) Closing the circle: sponsorship and the Greek Olympic games from ancient times to the present Day. The Int J History of Sport 22(4):618–638

Landström H, Parhankangas A, Mason C (2019) Handbook of research on crowdfunding. Edward Elgar, Cheltenham

Lovink G, Rossiter NM (2008) Reader: a critique of creative industries. Institute of Network Cultures, Amsterdam

Masters T (2013) Marillion 'understood where the internet was going early on', London, BBC. Accessed 1 Dec 2023 at https://www.bbc.com/news/entertainment-arts-23881382

McGuigan J (1966) Culture and the public sphere. Routledge, London

Mikić H (2015) A case study of Serbia: measuring the economic contribution of cultural industries, Montreal, UNESCO Institute for statistics

Mikić H (2022) Economic instruments and measures of cultural policy and fair practice: case of female artists and creative entrepreneurs in Serbia. In: Shabliy E, Engerman K, Kurochkin D (eds) Gender equity: global policies and perspectives on advancing social justice. London, Rowman & Littlefield, pp 39–57

O'Connor J (2016) After the Creative industries: cultural policy in crisis. Law, Soc Justice Global Developm 1
Pratt A (1997) The cultural industries production system: a case study of employment change in Britain, 1984–91. Environ Plan A 29(11):1953–1974
PwC (2023) PwC's global entertainment and media outlook 2023–2027, PWC, 2023. Accessed 1 Jan 2024. at https://www.pwc.com/gx/en/industries/tmt/media/outlook/insights-and-perspectives.html
Sattar D, Ramsden P (2000) Third sector financial instruments for the cultural sector. North West Arts Board, London
Statista (2023) In: Digital capital raising: market data and analysis, Hamburg, Statista
Throsby D (2006) An artistic production function: theory and an application to Australian visual artists. J Cult Econ 30:1–14
Tosatto J et al. (2019) An overview of crowdfunding in the creative and cultural industries. In: Landström H, Parhankangas A, Mason C (eds) Handbook of research on crowdfunding, London, Edward Elgar
UNCTAD (2022) Creative economy outlook 2022. UNCTAD, New York
UNESCO (1982) Mexico city declaration on cultural policies. UNESCO: World Conference on Cultural policy, Mexico City, 26 Julay-6 August 1982
Uspenski A (2022) TS Ventures investirao u domaće startape Anari AI, Bitebell i brush galaxy. Accessed 1 Dec 2023 at https://www.netokracija.rs/ts-ventures-investicije-202145
Ventu (2024) Financial teaser for film "Mudbrick", Accessed 1 Dec 2023. https://ventu.rs/campaigns/kuca-od-blata-mir-media/
Walldof J (2018) Mozart and finances. In: Keefe S (ed) Mozart in context. Cambridge University Press, Cambridge, pp 170–180
Zelenskaya E, Shakina E (2023) The efficiency of the performing arts: evidence from data envelopment analysis for Russian theaters. J Arts Manag Law Soc 53(5):273–294

Empowering Urban Resilience: Harnessing Crowdfunding for Sustainable Green Infrastructure

Ali Akbar Firoozi and Ali Asghar Firoozi

Abstract In an era characterized by rapid urbanization and growing climate change challenges, the need for resilient and sustainable urban environments has become critical. Green infrastructure, providing benefits such as enhanced biodiversity, flood risk mitigation, and improved public well-being, often faces financial constraints with traditional funding methods falling short. This chapter delves into crowdfunding as a promising solution that democratizes funding for green infrastructure, enhancing urban resilience. It analyzes various crowdfunding platforms and global case studies to illustrate how these participatory financing models not only raise essential funds but also foster significant community involvement in sustainable urban development. The discussion addresses major obstacles like regulatory issues and public perception that limit the effectiveness and growth of crowdfunding. Strategic recommendations, including policy changes, technological advancements, and community-focused approaches, are proposed to overcome these challenges. The chapter concludes by emphasizing the potential of new technologies and policy innovations to enhance the impact of green infrastructure projects. This thorough examination is designed to be a crucial resource for researchers, policymakers, urban planners, and community advocates, advocating for crowdfunding as a key transformative financing tool for sustainable urban resilience.

Keywords Crowdfunding · Green infrastructure · Urban resilience · Sustainable urban development · Community engagement · Innovative financing

A. A. Firoozi (✉)
Department of Civil Engineering, Faculty of Engineering and Technology, University of Botswana, Gaborone, Botswana
e-mail: a.firoozi@gmail.com

A. A. Firoozi
Department of Civil Engineering, Faculty of Engineering, National University of Malaysia (UKM), selangor, Malaysia

H. Baber and M. Fanea-Ivanovici (eds.), *Sustainable Financing—A Contemporary Guide for Green Finance, Crowdfunding and Digital Currencies*, World Sustainability Series, https://doi.org/10.1007/978-3-031-80969-9_13

1 Introduction

The exploration of innovative financing mechanisms for sustainable urban development, particularly through the lens of crowdfunding for green infrastructure (GI), marks a pivotal intersection of community engagement, environmental stewardship, and urban resilience. The inception of green infrastructure as a strategic and integrative approach to urban planning reflects a growing recognition of the need to harmonize ecological preservation with urban development, ensuring cities not only thrive but do so in a manner that is sustainable, resilient, and inclusive (Zhang et al. 2019; Pauleit et al. 2019).

Green infrastructure embodies the essence of sustainable urban development by weaving natural elements into the urban fabric, thus offering a counterbalance to the concrete jungle. It provides essential ecosystem services such as stormwater mitigation, air purification, and temperature regulation, which are crucial for the well-being of urban populations and the resilience of cityscapes against the adverse effects of climate change (Meerow et al. 2016; Pricope et al. 2019). These services are not just ancillary benefits but fundamental requisites for cities aiming to navigate the uncertain terrains of future climatic scenarios and urban growth dynamics.

However, the ambition to blanket cities with green infrastructure confronts a formidable obstacle: funding. Traditional financing mechanisms are often beleaguered by inconsistencies and insufficiencies, hampered by fluctuating political wills and competing budgetary demands (Sturiale and Scuderi 2019). This predicament has ushered in a growing interest in alternative financing solutions, among which crowdfunding has gained notable traction. Crowdfunding represents a paradigm shift in funding green infrastructure projects, enabling the direct involvement of the wider community in realizing urban sustainability ambitions (Buttice et al. 2017; Lam and Law 2016).

This participatory approach to financing does more than just fill the financial void; it cultivates a shared sense of purpose and collective ownership among urban residents towards their communal spaces. By leveraging the power of the crowd, crowdfunding not only gathers the requisite financial resources but also fosters a robust community engagement, ensuring that projects align with the social, environmental, and aesthetic values of the community (Jiao and Ma 2021).

This chapter, therefore, seeks to delve deeply into the dynamics of crowdfunding as a pivotal tool for fostering green infrastructure within urban landscapes. It endeavors to dissect the multifaceted roles that crowdfunding plays in bridging the financial gap, engaging communities, and steering urban development towards more sustainable and resilient futures. Through an analysis of select case studies, the chapter aims to illuminate the successful applications of crowdfunding in this domain, scrutinize the challenges inherent to its adoption, and distill critical lessons and strategies that can enhance its effectiveness. In essence, this inquiry contributes to the burgeoning dialogue on sustainable urban finance, charting a course for future research and practice aimed at leveraging crowdfunding for the ecological and social renewal of our cities.

2 Background

The exploration of crowdfunding as a viable avenue for financing green infrastructure (GI) initiatives is underpinned by a rich conceptual framework that navigates the confluence of environmental sustainability, community engagement, and financial innovation. At this intersection, green infrastructure emerges as a transformative approach to urban environmental management, ingeniously integrating natural processes within the urban fabric to address contemporary environmental challenges. This strategy enhances urban livability and socioeconomic vitality, heralding a new era of urban development characterized by resilience, sustainability, and participatory governance.

2.1 Defining Green Infrastructure

Green infrastructure is defined as an interconnected network of natural and semi-natural spaces that offers a comprehensive solution to urban environmental challenges. This innovative approach encompasses a diverse array of elements—including parks, green roofs, urban forests, greenways, and stormwater management systems—each playing a pivotal role in bolstering urban resilience and sustainability. The European Commission (Paulin et al. 2020) emphasizes green infrastructure's vital role in delivering key ecosystem services such as biodiversity conservation, climate regulation, and water management, positioning it as a foundational pillar of sustainable urban planning. Beyond its environmental contributions, green infrastructure fosters social cohesion, promotes mental and physical health, and drives economic growth by creating employment opportunities and enhancing property values. In the face of escalating climate change pressures, the adaptive and mitigative functions of green infrastructure are increasingly indispensable, charting a course toward more resilient and adaptable urban communities (Ramyar et al. 2021; Kumar et al. 2019).

2.2 Understanding Crowdfunding

Crowdfunding emerges as a transformative force in the domain of project financing, challenging traditional mechanisms by harnessing the collective power of the internet to support a wide spectrum of initiatives, including those aimed at entrepreneurial, social, and environmental goals. This financing model's adaptability across donation-based, reward-based, equity-based, and debt-based forms allows it to cater to a diverse range of project requirements and donor expectations, making it particularly suited to the unique challenges and opportunities presented by green infrastructure projects. Crowdfunding stands out as a vital instrument for bypassing conventional financial

Table 13.1 Types of crowdfunding models for green infrastructure financing

Crowdfunding model	Characteristics	Advantages	Challenges
Donation-based	Backers donate to projects without expectation of financial return	Accessible to a wide audience; fosters community spirit	Limited by donors' willingness to give without direct returns
Reward-based	Backers receive non-financial rewards based on contribution level	Incentivizes contributions; creative engagement with backers	Managing rewards can be logistically complex
Equity-based	Backers receive a stake in the project or venture	Attracts serious investors; aligns project and backer interests	Regulatory complexity; requires clear valuation
Debt-based	Backers lend money to the project with the expectation of repayment	Provides structured financing; clear terms of repayment	Interest obligations; risk of default

obstacles, facilitating public contributions that not only provide necessary capital but also foster a sense of community ownership and active participation in environmental endeavors. This model significantly enhances the visibility of green infrastructure projects, stimulating grassroots support and strengthening the bond between urban communities and the natural environment (Ryoba et al. 2020; Kim et al. 2020; Lin et al. 2020).

Table 13.1 provides a comparative overview of the different crowdfunding models that can be utilized for financing green infrastructure projects, highlighting their unique characteristics, advantages, and challenges. Donation-based crowdfunding, characterized by its appeal to community spirit and philanthropy, encourages wide participation by allowing backers to contribute without expecting financial returns. This model, while fostering a strong sense of communal support, is limited by its reliance on altruistic motivations, which may not always be sufficient to meet project funding goals. Reward-based crowdfunding, on the other hand, engages contributors by offering tangible, non-financial rewards in return for their support, effectively linking community contributions to specific project milestones. Despite its potential to incentivize contributions creatively, this model faces logistical challenges in managing and fulfilling reward commitments.

2.3 The Nexus Between Crowdfunding and Urban Resilience

The convergence of crowdfunding and urban resilience heralds a significant paradigm shift, spotlighting the empowering role of participatory finance in fostering community-driven urban development. This intersection is particularly poignant in the realm of green infrastructure, where crowdfunding not only catalyzes financial resources but also galvanizes community engagement, driving the collective effort towards creating resilient, sustainable urban landscapes. The unique advantage of crowdfunding lies in its ability to tap into a wider pool of resources, democratizing the funding process and enabling the execution of innovative projects that might otherwise be sidelined by conventional financing avenues. Moreover, this model of financing excels in engaging communities directly in resilience-building efforts, encouraging a participatory approach to urban planning that seamlessly integrates the principles of resilience into the fabric of city development. This collaborative endeavor significantly amplifies the impact of green infrastructure projects, ensuring that urban areas are better equipped to navigate the challenges posed by a rapidly changing climate and other future uncertainties (Junqueira et al. 2021; Dixon et al. 2018).

This nexus between green infrastructure and crowdfunding outlines a holistic strategy aimed at advancing urban development towards greater sustainability and resilience. By exploring this synergistic relationship, the aim is to illuminate how crowdfunding can act as a pivotal lever for the widespread implementation of green infrastructure initiatives, fostering environments that are not only environmentally sound but also conducive to enhancing the quality of urban life. Through such exploration, the potential of crowdfunding to revolutionize the approach to urban resilience becomes evident, promising a pathway towards cities that are more adaptable, robust, and inclusive.

Figure 13.1 serves as a visual representation of the cyclical, symbiotic relationship that crowdfunding fosters within the ecosystem of urban resilience. This diagram illustrates the dynamic process by which community-driven financial backing initiates and sustains the development of green infrastructure, underpinning a collaborative model of urban planning. The journey depicted—from the mobilization of public support and investment in green infrastructure to the tangible improvements in urban resilience, public health, and environmental quality—encapsulates the virtuous cycle engendered by participatory finance. Such a model not only underscores the direct benefits of community engagement in fostering urban resilience but also highlights the reinforcing feedback loop wherein enhanced urban spaces further motivate community involvement and support. This sustainable cycle of growth and resilience, driven by the collective efforts of urban communities, lays the foundation for a future where cities are not merely habitable but thrive as vibrant, resilient, and sustainable ecosystems.

3 Crowdfunding Platforms and Case Studies

The exploration of crowdfunding platforms and case studies reveals a dynamic landscape where financial innovation aligns with the goals of urban ecological and social resilience. This emerging narrative showcases the power of community-driven support in funding green infrastructure projects, highlighting platforms and initiatives

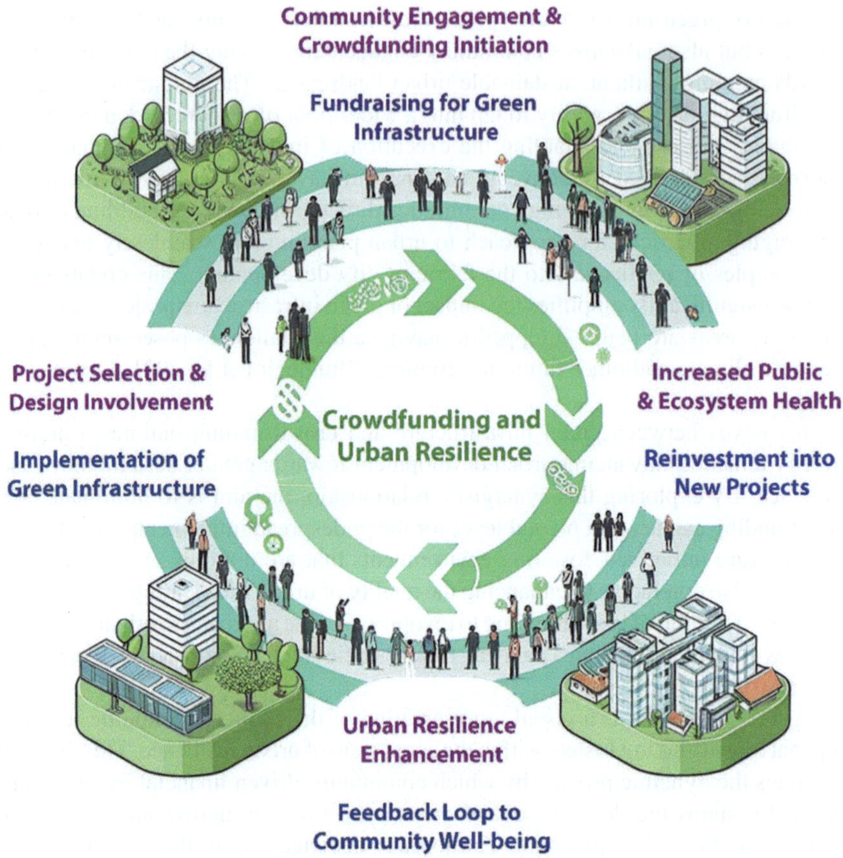

Fig. 13.1 Participatory growth cycle: crowdfunding for urban resilience

that have successfully bridged funding gaps and catalyzed significant environmental and societal advancements.

3.1 Overview of Major Crowdfunding Platforms Focused on Environmental Projects

The crowdfunding ecosystem is diverse, with platforms like Kickstarter and Indiegogo catalyzing numerous environmentally focused campaigns, demonstrating the versatility of crowdfunding in mobilizing resources for a wide array of projects. Meanwhile, specialized platforms such as GoFundMe have been instrumental in galvanizing community-led initiatives aimed at local environmental conservation and enhancement efforts. The emergence of platforms explicitly dedicated to sustainability projects, like Crowdfunder (UK), StartSomeGood, and GreenFunder, signals a growing dedication to environmental and social impact projects. These platforms carve out unique spaces where green infrastructure projects can connect with potential backers passionate about sustainability, offering not just financial support but also enhancing community engagement and awareness around such initiatives. Leveraging the expansive reach of social media and online networks, these platforms significantly amplify the visibility of green infrastructure projects, drawing in support from a global audience and marking a pivotal shift towards more collaborative, sustainable urban development practices (Anglin and Pidduck 2022; Manning and Bejarano 2017; Junge et al. 2022).

To further elucidate the intricacies of crowdfunding for green infrastructure, we will expand our analysis to include a more detailed examination of how these platforms have supported specific projects. This will involve a closer look at the mechanisms of engagement and funding, the challenges projects face in meeting their financial goals, and the strategies employed to overcome these challenges. Moreover, by analyzing the success factors behind thriving projects, we can provide actionable insights for future initiatives aiming to harness the power of crowdfunding for environmental sustainability.

3.2 Detailed Analysis of Successful Case Studies

A deep dive into successful case studies of crowdfunding for green infrastructure projects not only reveals their profound and varied impacts on urban landscapes but also brings to light the nuanced challenges and innovative solutions that have marked their journeys. These examples—the High Line in New York City, the Luchtsingel Bridge in Rotterdam, and the deployment of tree-planting drones in Myanmar—each stand out as beacons of innovation, community involvement, and environmental stewardship. Their stories enrich our understanding of how crowdfunding can catalyze sustainable urban development, offering lessons on overcoming obstacles and harnessing the collective power of communities.

1. *The High Line, New York City, USA*: Transforming an obsolete elevated railway into a thriving public park, the High Line project faced significant challenges, including regulatory hurdles, securing initial funding, and engaging a diverse urban community. Strategies to overcome these obstacles included extensive community outreach to build support, innovative fundraising events, and leveraging public–private partnerships. Key success factors were its visionary leadership, the compelling narrative of urban transformation, and strong community advocacy, which together secured its place as a landmark project (Popescul et al. 2020).
2. *The Luchtsingel Bridge, Rotterdam, Netherlands*: The Luchtsingel Bridge's crowdfunding campaign navigated challenges such as convincing a skeptical

public of the project's feasibility and managing the logistics of a participatory design process. The project team addressed these through transparent communication, engaging storytelling, and offering tangible rewards (personalized planks) for contributions. Its success was underpinned by its direct appeal to community pride, the innovative use of crowdfunding for public infrastructure, and the fostering of a sense of ownership among contributors, demonstrating the power of community resources in urban improvement (Bernardino and Santos 2020).

3. *Tree Planting Drone Technology, Myanmar*: Pioneering the use of drones for reforestation, this project confronted obstacles like demonstrating the technology's effectiveness and scalability and ensuring sustainable impact. Solutions included pilot projects to showcase the technology, engaging video content for crowdfunding campaigns, and partnerships with local organizations for ground support. The key to its success lay in leveraging cutting-edge technology to address environmental challenges, clear communication of its environmental impact, and community involvement in project execution, showcasing the potential for technology-driven solutions in environmental restoration (Golimowska 2016).

3.2.1 Challenges and Strategies Across Case Studies

Across these case studies, several recurring themes emerge in the challenges faced and strategies employed:

- *Regulatory Navigation*: Projects often encountered complex regulatory environments that required careful navigation. Strategies included engaging legal experts early and actively lobbying for regulatory support.
- *Community Engagement*: Each project demonstrated the critical importance of engaging the community not just as funders but as active participants in the project's development. Strategies involved extensive outreach, public workshops, and leveraging social media to build and maintain momentum.
- *Innovative Funding Approaches*: Facing the challenge of raising substantial funds, projects employed innovative approaches beyond traditional crowdfunding, such as matching funds from private entities, public–private partnerships, and offering tangible rewards to donors.
- *Sustainability and Impact Communication*: Ensuring and communicating the long-term sustainability and impact of the projects were pivotal. This involved clear, transparent reporting on project goals, progress, and outcomes, as well as establishing mechanisms for ongoing community involvement and stewardship.

3.2.2 Key Success Factors Identified

- *Strong Leadership and Clear Vision*: Essential for navigating challenges and maintaining project momentum.

- *Community Involvement and Ownership*: Projects succeeded by fostering a deep sense of community ownership and engagement.
- *Transparency and Effective Communication*: Crucial for building trust and sustaining support.
- *Innovative Solutions to Funding and Execution*: Including leveraging technology, creative fundraising strategies, and flexible project planning to adapt to unforeseen challenges.

Through the lens of these detailed case studies, we gain invaluable insights into the dynamics of crowdfunding for green infrastructure. These narratives not only celebrate the achievements but also serve as practical guides for overcoming the multifaceted challenges inherent in bringing such visionary projects to fruition. They underscore the synergy between community funding, participatory involvement, and sustainable development, reaffirming crowdfunding's role as a versatile and potent tool for advancing urban resilience and sustainability.

3.3 Lessons Learned and Impact on Local Communities

The journey through successful crowdfunding campaigns for green infrastructure projects not only unveils pivotal lessons but also significantly underscores the transformative impact these initiatives can wield on local communities. These insights, gleaned from a diverse array of projects, illuminate both the mechanics of crowdfunding and its potent capacity to foster sustainable urban environments.

- *Deepened Community Engagement and Enhanced Ownership*: A cornerstone of crowdfunding's success in green infrastructure projects lies in the robust engagement and sense of ownership it fosters within local communities. The detailed case studies illuminate how projects that actively involve communities from their inception—not merely as donors but as co-creators and stewards—dramatically increase their chances of success and enduring impact. This participatory approach deepens the connection residents have with their urban environment, fostering a culture of care and a vested interest in the sustained health of green spaces. Projects that excel in this regard often employ innovative engagement strategies, such as participatory planning workshops, digital engagement tools, and public forums, ensuring the community's voice and preferences shape project outcomes (Lang and Rothenberg 2017; Sedlitzky and Franz 2019).
- *Innovation in Funding Mechanisms*: Crowdfunding has emerged as a groundbreaking funding mechanism that effectively democratizes the process of capital mobilization for public goods. It challenges and transcends traditional financial barriers, enabling the realization of projects that, despite their significant social and environmental merits, may not meet conventional investment criteria.

This inclusive financing model allows broad participation, empowering individuals to contribute to their community's betterment. By exploring diverse crowdfunding models—from donation-based to equity-based platforms—projects have unlocked novel avenues for financing, illustrating the versatility and adaptability of crowdfunding in various urban contexts (Passeri 2020; Fish 2023).

- *Clarity, Communication, and Transparency*: The bedrock of a successful crowdfunding campaign lies in its ability to communicate transparently and effectively with potential backers. Projects that articulate their objectives, anticipated outcomes, and the environmental and social benefits of the proposed green infrastructure clearly and compellingly tend to attract more support. This communication strategy not only builds trust among community members and potential investors but also ensures projects align with community needs and expectations. The case studies highlight the importance of leveraging multiple communication channels, including social media, community meetings, and local media, to maintain dialogue and transparency throughout the project's lifecycle (Cruzan et al. 2016).
- *Multifaceted Community Impacts*: The impacts of crowdfunding on local communities extend far beyond the tangible enhancements of urban landscapes. They also encompass intangible benefits, such as fostering community cohesion, enhancing environmental awareness, and contributing to social well-being. Successful projects catalyze economic development, create employment opportunities, invigorate local economies, and improve public health by providing cleaner, greener, and more resilient urban spaces. These projects serve as exemplars of how crowdfunding can act as a catalyst for comprehensive social change, reinforcing the importance of community solidarity and collective action in urban development (Zhou et al. 2018).
- *Contributions to Urban Resilience and Sustainability*: Crowdfunding initiatives for green infrastructure contribute significantly to the long-term resilience and sustainability of urban communities. By addressing both immediate and future funding needs, these projects lay the groundwork for more adaptable, robust, and thriving urban environments. They underscore the necessity of viewing green infrastructure investments not as expenses but as critical components of a city's resilience strategy, offering a blueprint for future efforts in urban sustainability.

In synthesizing these lessons and impacts, it becomes evident that participatory finance, through crowdfunding, offers a powerful means to develop and implement green infrastructure projects that are both impactful and sustainable. By prioritizing community engagement, embracing innovative funding models, and upholding transparency and clear communication, stakeholders can harness crowdfunding's full potential. These revised insights aim to provide comprehensive guidance for future projects, charting a path toward creating urban environments that are not only livable but flourish as resilient and sustainable ecosystems for generations to come.

4 Challenges and Opportunities

Navigating the terrain of crowdfunding for green infrastructure projects unveils a dual landscape of significant challenges and promising opportunities. This intricate interplay embodies the essence of innovation in financial strategies tailored to sustainable urban development, where obstacles serve as catalysts for strategic advancements and creative solutions. At the heart of this journey lies a commitment to community empowerment and the overarching goals of resilience and sustainability, positioning crowdfunding not merely as a tool for financial mobilization but as a foundational pillar for participatory and inclusive urban development.

4.1 Challenges in Crowdfunding for Green Infrastructure

The path to realizing green infrastructure projects through crowdfunding is fraught with challenges that test the limits of this financing model's efficacy and scalability:

1. *Regulatory Hurdles*: The crowdfunding arena is mired in a complex web of legal and regulatory frameworks that differ markedly across jurisdictions. These regulatory challenges, particularly pronounced for equity crowdfunding and public investments, demand rigorous compliance to uphold the campaigns' legality and integrity. Such demands can impose considerable administrative and financial strains on project initiators, potentially stifling innovation and deterring participation in crowdfunding ventures for green infrastructure (Zhou et al. 2018).
2. *Public Perception and Awareness*: The success of a crowdfunding campaign hinges critically on its ability to engage a wide audience and articulate the project's value convincingly. For green infrastructure initiatives, overcoming public skepticism and fostering trust are paramount. This pivotal challenge necessitates transparent communication and robust public outreach to enhance awareness of the projects' potential contributions to urban resilience and environmental sustainability (Cecere et al. 2017).
3. *Project Viability and Execution*: Ensuring the technical and financial feasibility of green infrastructure projects stands as a formidable barrier. Crowdfunding campaigns are tasked with demonstrating that their proposed initiatives can achieve stated objectives within the confines of the allocated timeline and budget, all while delivering anticipated environmental and social benefits. Addressing this challenge requires comprehensive planning, meticulous feasibility studies, and the engagement of multidisciplinary experts (Dalla Chiesa and Handke 2020).
4. *Sustainability of Funding*: While crowdfunding can successfully secure the initial capital needed for green infrastructure projects, guaranteeing their long-term sustainability and operational viability presents an ongoing financial challenge. This issue accentuates the necessity for innovative funding sources or revenue-generating mechanisms to ensure the continuous impact and maintenance of the projects over time (Parker and Simpson 2020).

Table 13.2 Summary of challenges in crowdfunding for green infrastructure

Challenge	Description	Potential impact on crowdfunding initiatives
Regulatory hurdles	Navigating complex legal and regulatory environments across jurisdictions	Creates uncertainty, increases administrative and financial burdens
Public perception and awareness	Overcoming skepticism and building trust among potential backers	Affects the ability to engage a broad audience and secure funding
Project viability and execution	Demonstrating the technical and financial feasibility of projects	Impacts the credibility and attractiveness of crowdfunding campaigns
Sustainability of funding	Ensuring the long-term financial sustainability of projects beyond initial capital collection	Challenges the maintenance and operational longevity of funded projects

Table 13.2 distills the primary challenges faced by crowdfunding initiatives within the green infrastructure domain, offering insights into the potential hurdles that may impede the success and scalability of these ventures. Each challenge outlined in the table underscores a critical aspect of crowdfunding's complex landscape, from navigating regulatory intricacies and building public trust to ensuring project viability and long-term sustainability.

Figure 13.2, the Crowdfunding Challenge Cycle, visually encapsulates the myriad challenges crowdfunding initiatives encounter. This diagram illustrates the dynamic, cyclical nature of overcoming obstacles in the crowdfunding ecosystem, highlighting the iterative process of navigating regulatory adaptation, addressing perceptual barriers, ensuring project viability, and achieving sustainable growth. Central to this depiction is the Challenge Response Loop, emphasizing the continuous evolution and adaptation required to surmount these hurdles effectively. This visual representation aids in comprehending the multifaceted ecosystem within which crowdfunding operates, underscoring the imperative for resilience, creativity, and strategic planning in leveraging crowdfunding for the development of resilient and sustainable urban spaces.

4.2 *Opportunities for Overcoming These Challenges*

Despite the array of challenges that confront crowdfunding in the realm of green infrastructure, a landscape of opportunities exists that can significantly bolster the effectiveness of such campaigns. These opportunities not only offer pathways to surmount existing hurdles but also serve to amplify the impact of crowdfunding efforts, driving forward the development of resilient and sustainable urban environments.

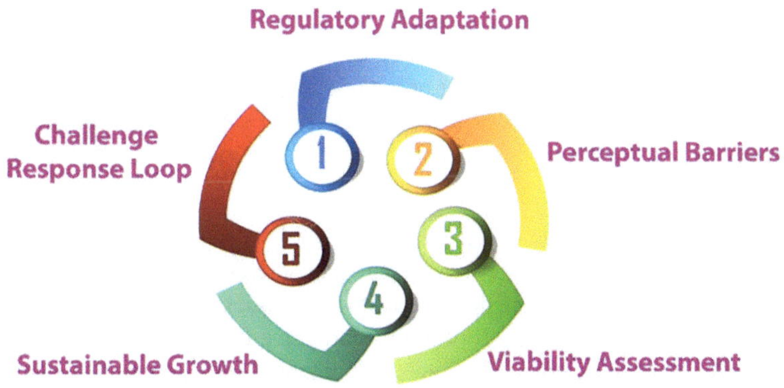

Fig. 13.2 The crowdfunding challenge cycle

1. *Leveraging Digital Platforms*: The digital era presents unprecedented opportunities to enhance the visibility and reach of crowdfunding campaigns. Utilizing social media and digital marketing strategies can exponentially increase a campaign's exposure, attracting a diverse audience and fostering a robust sense of community engagement around green infrastructure projects. By tapping into the collective power of online platforms, crowdfunding initiatives can transcend geographical boundaries, uniting a global community of supporters around shared environmental goals (Khirfan and El-Shayeb 2020).
2. *Partnerships and Collaborations*: The strength of crowdfunding is magnified through strategic partnerships and collaborations. By aligning with local governments, non-profit organizations, and businesses, crowdfunding campaigns gain additional resources, expertise, and credibility. These alliances are instrumental in navigating the complexities of regulatory landscapes and in bolstering the long-term viability of green infrastructure projects. Collaborative efforts not only lend projects a greater sense of legitimacy but also pool a wider array of resources and knowledge, critical for the successful implementation and sustainability of projects (Tan and Reddy 2021).
3. *Innovative Reward Systems*: A compelling rewards system can serve as a powerful incentive for potential backers, transforming the crowdfunding experience from a mere financial transaction to a meaningful contribution towards a sustainable future. By offering rewards that resonate with the community's values and aspirations—ranging from recognition in project materials to opportunities for active participation in the development process—campaigns can significantly enhance contributor engagement and motivation. Such innovative reward schemes encourage a deeper investment in the success of green infrastructure projects, aligning backers' contributions with tangible outcomes and benefits (Hehn and Uebernickel 2018).
4. *Education and Awareness Campaigns*: At the core of successful crowdfunding efforts lies the capacity to effectively communicate the importance and impact of green infrastructure projects. Targeted education and awareness campaigns

play a pivotal role in building robust public support, dispelling misconceptions, and fostering an informed community of backers. By clearly articulating the benefits of green infrastructure for urban resilience, sustainability, and quality of life, such campaigns can cultivate a groundswell of support, transforming public perception and fostering a culture of environmental stewardship (Hudson et al. 2016).

Table 13.3 aligns the identified challenges with corresponding opportunities, presenting a strategic framework for overcoming obstacles in crowdfunding for green infrastructure. This table delineates actionable strategies, from policy reform and community engagement to the adoption of innovative financing models, providing a blueprint for leveraging crowdfunding's full potential in the pursuit of sustainable urban development.

Figure 13.3 encapsulates the dynamic ecosystem of crowdfunding, illustrating the cyclical process of overcoming challenges through regulatory adaptation, community engagement, project viability assessments, and the pursuit of sustainable growth. Central to this diagram is the Challenge Response Loop, highlighting the iterative and responsive nature of successful crowdfunding initiatives. This visual representation underscores the importance of flexibility, innovation, and resilience in navigating the crowdfunding landscape, paving the way for the realization of green infrastructure projects that contribute to the resilience and sustainability of urban environments.

Table 13.3 Opportunities for overcoming crowdfunding challenges

Challenge	Opportunity	Strategy description
Regulatory hurdles	Simplification and harmonization of regulations	Advocate for policy reforms that simplify crowdfunding regulations and create a more harmonious legal framework across jurisdictions to reduce administrative burdens
Public perception and awareness	Targeted education and marketing campaigns	Implement comprehensive education and awareness campaigns that accurately convey the benefits and viability of green infrastructure projects to build trust and public support
Project viability and execution	Expert involvement and feasibility studies	Engage experts in early project planning stages and conduct thorough feasibility studies to ensure projects are technically and financially viable
Sustainability of funding	Innovative financing models	Explore and adopt innovative financing models, such as matching funds, grants, or revenue-generating activities, to ensure the long-term sustainability of projects

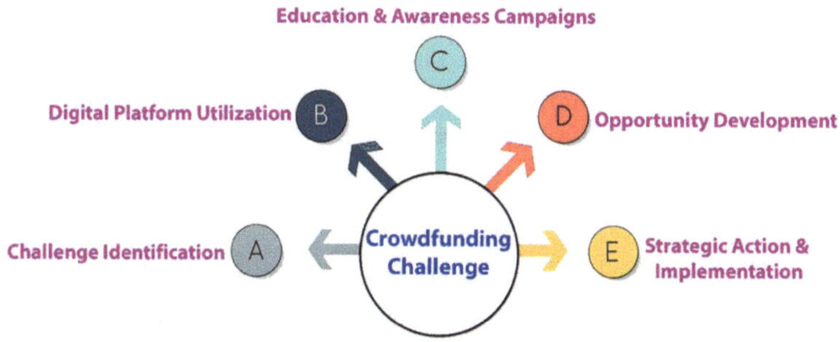

Fig. 13.3 Ecosystem of crowdfunding

4.3 *Ensuring Long-Term Sustainability and Resilience of Crowdfunded Projects*

In the realm of crowdfunding for green infrastructure, the journey from conceptualization to realization is fraught with challenges, yet it is the quest for long-term sustainability and resilience that often proves most daunting. Achieving enduring success necessitates a multifaceted approach, emphasizing the importance of monitoring, evaluation, community involvement, and adaptive management.

Monitoring and Evaluation (M&E) Frameworks are indispensable for assessing the ongoing impact of projects. A well-designed M&E framework offers a systematic method to measure environmental, social, and economic outcomes, enabling project managers to track progress, identify areas for improvement, and demonstrate the project's benefits to stakeholders and funders. For instance, an urban park revitalization project might utilize an M&E framework to document increases in biodiversity, enhancements in air and water quality, and community utilization patterns, providing valuable insights that can guide future urban greening initiatives.

Sustainability Planning forms the cornerstone of any long-term project strategy. From the outset, projects must consider not only the initial implementation but also future maintenance, environmental adaptability, and community value. This involves integrating sustainable practices, such as the use of renewable energy sources for maintenance operations and designing with climate resilience in mind. Additionally, establishing mechanisms for ongoing funding, whether through endowments, grants, or continuous crowdfunding efforts, ensures projects remain financially viable and continue to serve community needs.

The role of Community Stewardship cannot be overstated in extending the life and relevance of green infrastructure projects. By fostering deep community connections and ownership through stewardship programs, projects can tap into local knowledge and enthusiasm, ensuring that green spaces are cared for and cherished by those who use them most. These programs not only enhance the project's sustainability but also enrich the social fabric of the community, creating a virtuous cycle of engagement and benefit.

Adaptive Management offers a pragmatic approach to navigating the uncertainties of environmental change and urban development. By remaining flexible and responsive to new challenges and information, projects can evolve over time, adjusting strategies to ensure ongoing relevance and impact. This might include modifying planting techniques in response to survival rates or evolving designs to accommodate new urban pressures, ensuring that projects continue to meet their environmental and social objectives.

Finally, securing Policy and Regulatory Support is critical for embedding green infrastructure projects within the larger urban planning and resilience frameworks. Advocacy for supportive policies and regulations can provide a more stable foundation for these projects, ensuring they are not only protected but prioritized within urban development agendas.

In conclusion, the path to ensuring the long-term sustainability and resilience of crowdfunded green infrastructure projects is complex and requires a proactive, comprehensive strategy. By embracing monitoring and evaluation, forward-thinking sustainability planning, community stewardship, adaptive management, and policy advocacy, these projects can achieve lasting impact, contributing to the creation of more resilient, vibrant, and sustainable urban environments. Through such efforts, crowdfunded green infrastructure projects not only enhance the present-day urban landscape but also lay the groundwork for future generations to enjoy and cherish.

5 Engaging Communities in Green Infrastructure Projects

The essence of successful green infrastructure projects lies in the profound engagement and collaboration of communities. This section delves into the various strategies that can significantly enhance community involvement, ensuring that green infrastructure projects are not only environmentally impactful but also deeply rooted in the social and cultural tapestry of the community.

5.1 The Importance of Community Engagement

The importance of community engagement in the development and success of green infrastructure projects cannot be overstated. This engagement goes beyond simple participation, fostering a profound connection between individuals and their urban environments. Such deep-rooted involvement is instrumental in promoting a collective sense of ownership, responsibility, and stewardship over local ecosystems and communal spaces. When community members view themselves as active stewards of their environment, they are more inclined to contribute positively to its preservation and enhancement, ensuring the sustainability and success of green infrastructure initiatives. This sense of belonging and collective stewardship not only bolsters the acceptance and longevity of these projects but also ensures their effective integration into the urban fabric, making them more responsive to the community's needs and values (Foster and Iaione 2019).

Active community participation in the maintenance and stewardship of green infrastructure projects is a testament to the transformative power of community engagement. By involving local populations in the care of green spaces, cities can cultivate a stronger sense of ownership among residents, making them key stakeholders in the success and longevity of these projects. This involvement can take various forms, from volunteer days for planting native vegetation to community-led workshops on sustainable water use, enhancing both the physical landscape and the community's connection to their environment (Anguelovski 2016).

At the core of effective green infrastructure projects is their ability to meet the specific needs and preferences of the local population, best achieved through participatory planning processes that engage community members from the outset. This collaborative approach ensures that the resulting green spaces are not only functional and sustainable but also resonant with the community's identity and values, leading to higher rates of use and engagement (Heynen et al. 2018).

The direct involvement of communities in the design and implementation of green infrastructure projects significantly boosts their acceptance and utilization. When individuals see their ideas and feedback reflected in the development of urban green spaces, they are more likely to support and use these areas. This heightened sense of community buy-in is crucial for the long-term success and impact of green infrastructure, leading to spaces that are not only environmentally beneficial but also cherished and actively used by the local population (Mell 2017).

In conclusion, the integration of community engagement at every stage of green infrastructure projects, from planning to maintenance, is vital. It ensures that these projects not only achieve their environmental goals but also strengthen community ties and improve the quality of urban life. Through fostering a sense of ownership, addressing community needs, and enhancing project acceptance, community engagement emerges as a cornerstone of sustainable urban development. By prioritizing the involvement of local populations, cities can unlock the full potential of green infrastructure to create more resilient, vibrant, and inclusive urban spaces, underscoring the indispensable role of community members in transforming urban environments into sustainable and livable habitats for current and future generations (Kahne et al. 2016).

5.2 Strategies for Effective Communication and Stakeholder Involvement

The key to fostering meaningful community engagement in green infrastructure projects is adopting a multifaceted approach that combines traditional methods with innovative strategies to reach and involve as broad an audience as possible.

- *Public Workshops and Consultations*: These sessions are invaluable for incorporating community input directly into project planning and execution, fostering a sense of ownership and commitment among residents. The High Line project in New York City is a stellar example, where community workshops played a crucial role in shaping the design and functionality of the park, ultimately contributing to its success and widespread acceptance (See et al. 2016).
- *Digital Engagement Platforms*: In today's digital age, leveraging social media and online forums can dramatically expand a project's reach, engaging a wider and more diverse audience. This strategy proved effective for the Luchtsingel Bridge in Rotterdam, where an online campaign facilitated crowdfunding and community voting, playing a pivotal role in the project's realization (Pera 2020).
- *Innovative Engagement Techniques*: Employing creative methods like participatory mapping, design competitions, and virtual reality simulations can make the engagement process more interactive and accessible. For instance, the Tree Planting Drone Technology project in Myanmar utilized VR to visualize potential environmental impacts, fostering a deeper understanding and support among the community and potential backers (Zhao et al. 2022).
- *Partnerships with Local Organizations*: Collaborating with local schools, nonprofits, and business associations can significantly amplify a project's impact, ensuring a more inclusive and representative process of community engagement. The Detroit Water and Sewerage Department's Rain Gardens project exemplifies this approach, where partnerships with local educational institutions enhanced community involvement and education on sustainable water management practices (Anguelovski et al. 2022).

Figure 13.4 visually represents the spectrum of community engagement techniques in action, using icons and photos to illustrate the dynamic and participatory nature of these methods. From the tangible involvement in public workshops and the digital interaction on social media platforms to the innovative experiences offered by VR simulations, this figure captures the essence of community engagement in the modern urban context. It serves as an inspirational guide, showcasing the various ways communities can be actively involved in shaping their urban environments through green infrastructure projects.

Through the implementation of these strategies, green infrastructure projects can achieve greater success and sustainability, firmly anchored in the support and involvement of the community. By fostering a deep sense of ownership, encouraging active participation, and ensuring projects are attuned to the community's needs and values, urban environments can be transformed into more resilient, vibrant, and inclusive spaces that are cherished by their inhabitants.

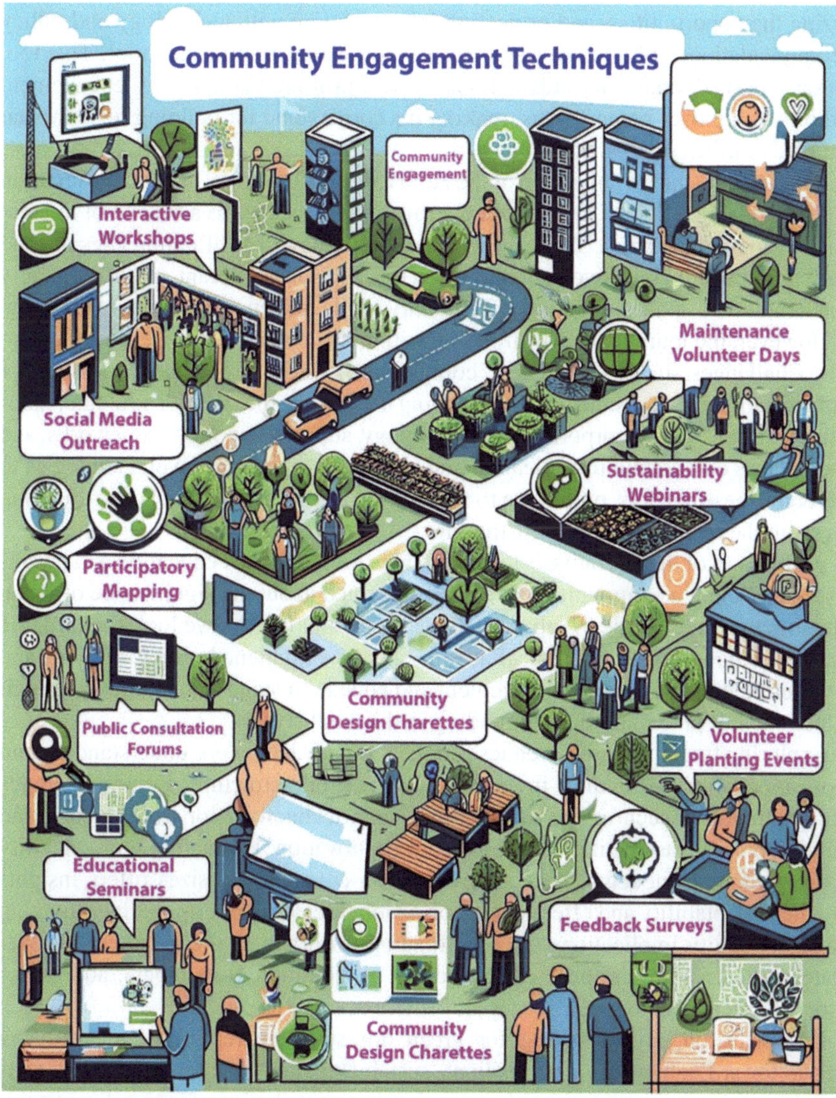

Fig. 13.4 Community engagement techniques in action

6 Conclusion

This chapter has embarked on an enlightening journey through the realm of crowdfunding for green infrastructure, unveiling its transformative potential as a key financier of sustainable urban development. By weaving through the conceptual framework, dissecting an array of successful case studies, navigating the challenges, seizing the opportunities, and projecting into the future with technology and policy recommendations, it has showcased crowdfunding as a multifaceted tool. This tool not only bridges financial voids but also fosters profound community engagement and ownership over green spaces, making it a cornerstone for building resilient and sustainable urban ecosystems.

Illustrated by the revitalization of the High Line in New York City and the creation of the Luchtsingel Bridge in Rotterdam, crowdfunding shines as a beacon of innovation and community spirit. These projects exemplify how crowdfunding goes beyond funding to inspire collective action towards the creation of vibrant and ecologically beneficial urban spaces, enhancing the quality of life and environmental well-being.

However, the journey of crowdfunding within this domain is not devoid of obstacles. Challenges such as regulatory complexities, public skepticism, questions of project viability, and sustainability of funding underscore the need for strategic, innovative solutions. The incorporation of technology, supportive policy frameworks, and an emphasis on community education and awareness emerge as pivotal strategies to overcome these barriers, promising to amplify the reach and impact of crowdfunding efforts in the realm of green infrastructure.

At the heart of this narrative is the undeniable importance of education and awareness. Cultivating a community that is well-informed about the significance of green infrastructure and its role in urban resilience is paramount. As we look forward, the synergistic power of technological advancements, enlightened policy support, and unwavering efforts in public engagement and education stand as the pillars that will unlock the full potential of crowdfunding in this vital sector.

Exploring future research avenues is crucial for advancing our understanding and application of crowdfunding in supporting green infrastructure. Emerging trends in crowdfunding, such as the use of blockchain technology for transparency and security, present fascinating opportunities for innovation. Evaluating the scalability of crowdfunding models across different urban contexts and sizes offers insights into their adaptability and effectiveness on a global scale. Furthermore, investigating the social equity implications of crowdfunding for urban resilience could uncover strategies to ensure that these efforts benefit all community members, addressing disparities in access to green spaces and environmental resources.

In conclusion, crowdfunding for green infrastructure represents a promising avenue toward achieving more resilient, sustainable, and livable urban futures. Its capacity to engage communities, leverage modest contributions for significant environmental and social returns, and promote participatory urban development positions it as an invaluable resource in the urban planner's toolkit. As cities worldwide confront the dual challenges of climate change and urbanization, innovative financing models like crowdfunding will undoubtedly play a crucial role in shaping the future of urban resilience.

The journey of crowdfunding in green infrastructure is an evolving narrative, enriched by each project's achievements and lessons learned. It symbolizes a collective quest toward sustainable and inclusive urban futures, calling for the combined efforts of communities, policymakers, entrepreneurs, and academics. Together, we are invited to embrace this journey, forging resilient and vibrant cities for the generations to come, while continuously exploring new research paths that will illuminate the way forward.

References

Anglin AH, Pidduck RJ (2022) Choose your words carefully: harnessing the language of crowdfunding for success. Bus Horiz 65(1):43–58. https://doi.org/10.1016/j.bushor.2021.09.004
Anguelovski I (2016) From toxic sites to parks as (green) LULUs? New challenges of inequity, privilege, gentrification, and exclusion for urban environmental justice. J Plan Lit 31(1):23–36. https://doi.org/10.1177/0885412215610491
Anguelovski I, Brand AL, Ranganathan M, Hyra D (2022) Decolonizing the green city: from environmental privilege to emancipatory green justice. Environ Justice 15(1):1–11. https://doi.org/10.1089/env.2021.0014
Bernardino S, Santos JF (2020) Crowdfunding: an exploratory study on knowledge, benefits and barriers perceived by young potential entrepreneurs. J Risk and Finan Managem 13(4):81. https://doi.org/10.3390/jrfm13040081
Buttice V, Colombo MG, Wright M (2017) Serial crowdfunding, social capital, and project success. Entrep Theory Pract 41(2):183–207. https://doi.org/10.1111/etap.12271
Cecere G, Le Guel F, Rochelandet F (2017) Crowdfunding and social influence: an empirical investigation. Appl Econ 49(57):5802–5813. https://doi.org/10.1080/00036846.2017.1343450
Cruzan MB, Weinstein BG, Grasty MR, Kohrn BF, Hendrickson EC, Arredondo TM, Thompson PG (2016) Small unmanned aerial vehicles (micro-UAVs, drones) in plant ecology. Appl Plant Sci 4(9):1600041. https://doi.org/10.3732/apps.1600041
Dalla Chiesa C, Handke C (2020) Crowdfunding. In: Handbook of cultural economics. 3rd ed. Edward Elgar Publishing, pp 158–167. https://doi.org/10.4337/9781788975803.00023
Dixon T, Connaughton J, Green S (2018) Understanding and shaping sustainable futures in the built environment to 2050. In: Sustainable futures in the built environment to 2050: a foresight approach to construction and development. pp 337–364. https://doi.org/10.1002/9781119063834.ch17
Fish A (2023) Reforesting native America with drones: rooting carbon with arborescent governmentality and decolonial geoengineering. Theory Cult Soc 40(7–8):157–177. https://doi.org/10.1177/02632764221096815
Foster SR, Iaione C (2019) Ostrom in the city: design principles and practices for the urban commons. In: Routledge handbook of the study of the commons. Routledge, pp 235–255
Golimowska K (2016) The Post-9/11 city in novels: literary remappings of New York and London. McFarland
Hehn J, Uebernickel F (2018) The use of design thinking for requirements engineering: an ongoing case study in the field of innovative software-intensive systems. In: 2018 IEEE 26th international requirements engineering conference (RE), August, IEEE, pp 400–405. https://doi.org/10.1109/RE.2018.00-18
Heynen N, Aiello D, Keegan C, Luke N (2018) The enduring struggle for social justice and the city. Ann Am Assoc Geogr 108(2):301–316. https://doi.org/10.1080/24694452.2017.1419414
Hudson S, Huang L, Roth MS, Madden TJ (2016) The influence of social media interactions on consumer–brand relationships: a three-country study of brand perceptions and marketing behaviors. Int J Res Mark 33(1):27–41. https://doi.org/10.1016/j.ijresmar.2015.06.004
Jiao H, Ma L (2021) Why do people support online crowdfunding charities? A case study from China. Front Psychol 12:582508. https://doi.org/10.3389/fpsyg.2021.582508

Junge LB, Laursen IC, Nielsen KR (2022) Choosing crowdfunding: why do entrepreneurs choose to engage in crowdfunding? Technovation 111:102385. https://doi.org/10.1016/j.technovation.2021.102385

Junqueira JR, Serrao-Neumann S, White I (2021) Managing urban climate change risks: prospects for using green infrastructure to increase urban resilience to floods. In: The impacts of climate change. Elsevier, pp 379–396. https://doi.org/10.1016/B978-0-12-822373-4.00013-6

Kahne J, Hodgin E, Eidman-Aadahl E (2016) Redesigning civic education for the digital age: participatory politics and the pursuit of democratic engagement. Theory and Res Soc Educ 44(1):1–35. https://doi.org/10.1080/00933104.2015.1132646

Khirfan L, El-Shayeb H (2020) Urban climate resilience through socio-ecological planning: a case study in Charlottetown, Prince Edward Island. J Urbanism: Int Res Placemaking and Urban Sustain 13(2):187–212. https://doi.org/10.1080/17549175.2019.1650801

Kim C, Kannan PK, Trusov M, Ordanini A (2020) Modeling dynamics in crowdfunding. Mark Sci 39(2):339–365. https://doi.org/10.1287/mksc.2019.1209

Kumar P, Druckman A, Gallagher J, Gatersleben B, Allison S, Eisenman TS, Morawska L (2019) The nexus between air pollution, green infrastructure and human health. Environ Int 133:105181. https://doi.org/10.1016/j.envint.2019.105181

Lam PT, Law AO (2016) Crowdfunding for renewable and sustainable energy projects: an exploratory case study approach. Renew Sustain Energy Rev 60:11–20. https://doi.org/10.1016/j.rser.2016.01.046

Lang S, Rothenberg J (2017) Neoliberal urbanism, public space, and the greening of the growth machine: New York City's high line park. Environ Planning a: Econ Space 49(8):1743–1761. https://doi.org/10.1177/0308518X16677969

Lin WS, Chen HR, Yang ML (2020) How do we learn to get success together through crowdfunding platform? From the perspectives of system learning and multi-motivations. Telematics Inform 52:101428. https://doi.org/10.1016/j.tele.2020.101428

Manning S, Bejarano TA (2017) Convincing the crowd: entrepreneurial storytelling in crowdfunding campaigns. Strateg Organ 15(2):194–219. https://doi.org/10.1177/1476127016648500

Meerow S, Newell JP, Stults M (2016) Defining urban resilience: a review. Landsc Urban Plan 147:38–49. https://doi.org/10.1016/j.landurbplan.2015.11.011

Mell IC (2017) Green infrastructure: reflections on past, present and future praxis. Landsc Res 42(2):135–145. https://doi.org/10.1080/01426397.2016.1250875

Parker J, Simpson GD (2020) A theoretical framework for bolstering human-nature connections and urban resilience via green infrastructure. Land 9(8):252. https://doi.org/10.3390/land9080252

Passeri F (2020) Lessons learned from civic crowdfunding and match-funding schemes in Europe. In: Crowdasset: crowdfunding for policymakers. pp 211–255. https://doi.org/10.1142/9789811207822_0010

Pauleit S, Ambrose-Oji B, Andersson E, Anton B, Buijs A, Haase D, van den Bosch CK (2019) Advancing urban green infrastructure in Europe: outcomes and reflections from the GREEN SURGE project. Urban Forestry and Urban Greening 40:4–16. https://doi.org/10.1016/j.ufug.2018.10.006

Paulin MJ, Remme RP, de Nijs T, Rutgers M, Koopman KR, de Knegt B, Breure AM (2020) Application of the natural capital model to assess changes in ecosystem services from changes in green infrastructure in Amsterdam. Ecosyst Serv 43:101114. https://doi.org/10.1016/j.ecoser.2020.101114

Pera M (2020) Potential benefits and challenges of the relationship between social movements and the commons in the city of Barcelona. Ecol Econ 174:106670. https://doi.org/10.1016/j.ecolecon.2020.106670

Popescul D, Radu LD, Păvăloaia VD, Georgescu MR (2020) Psychological determinants of investor motivation in social media-based crowdfunding projects: a systematic review. Front Psychol 11:588121. https://doi.org/10.3389/fpsyg.2020.588121

Pricope NG, Mapes KL, Woodward KD (2019) Remote sensing of human–environment interactions in global change research: a review of advances, challenges and future directions. Remote Sensing 11(23):2783.https://doi.org/10.3390/rs11232783
Ramyar R, Ackerman A, Johnston DM (2021) Adapting cities for climate change through urban green infrastructure planning. Cities 117:103316. https://doi.org/10.1016/j.cities.2021.103316
Ryoba MJ, Qu S, Ji Y, Qu D (2020) The right time for crowd communication during campaigns for sustainable success of crowdfunding: evidence from Kickstarter platform. Sustainability 12(18):7642. https://doi.org/10.3390/su12187642
Sedlitzky R, Franz Y (2019) What if we all chip in?' Civic crowdfunding as alternative financing for urban development projects. Built Environ 45(1):26–44. https://doi.org/10.2148/benv.45.1.26
See L, Mooney P, Foody G, Bastin L, Comber A, Estima J, Rutzinger M (2016) Crowdsourcing, citizen science or volunteered geographic information? The current state of crowdsourced geographic information. ISPRS Int J Geo Inf 5(5):55. https://doi.org/10.3390/ijgi5050055
Sturiale L, Scuderi A (2019) The role of green infrastructures in urban planning for climate change adaptation. Climate 7(10):119. https://doi.org/10.3390/cli7100119
Tan YH, Reddy SK (2021) Crowdfunding digital platforms: backer networks and their impact on project outcomes. Soc Netw 64:158–172. https://doi.org/10.1016/j.socnet.2020.09.005
Zhang Z, Meerow S, Newell JP, Lindquist M (2019) Enhancing landscape connectivity through multifunctional green infrastructure corridor modelling and design. Urban Forestry and Urban Greening 38:305–317. https://doi.org/10.1016/j.ufug.2018.10.014
Zhao Y, van den Berg PE, Ossokina IV, Arentze TA (2022) Individual momentary experiences of neighbourhood public spaces: results of a virtual environment based stated preference experiment. Sustainability 14(9):4938. https://doi.org/10.3390/ijgi5050055
Zhou M, Lu B, Fan W, Wang GA (2018) Project description and crowdfunding success: an exploratory study. Inf Syst Front 20:259–274. https://doi.org/10.1007/s10796-016-9723-1

100 Quotes About Central Bank Digital Currencies

Peterson K. Ozili

Abstract The objective of this article is to present some of the current thinking and arguments about central bank digital currency (CBDC) from the perspective of those who support the issuance of a central bank digital currency (CBDC) and from those who are opposed to CBDC. The article gives the reader an opportunity to reflect deeply about CBDC and to make their own opinion about CBDC based on the informed insights of others. From the collection of quotes, it was found that the concept of a central bank digital currency has come to stay, and many central banks want to issue a CBDC in the distant future. It was also found that, despite the efforts of central banks and pro-CBDC enthusiasts to publicise the benefits of a CBDC for citizens, many people continue to raise strong concerns about the potential for government overreach and surveillance, loss of competitive advantage for deposit-taking financial institutions, loss of privacy for citizens, and concerns that CBDC development is an unwholesome distraction for central banks, among other concerns. There is also a perceived negative sentiment about CBDC, and this sentiment is unlikely to change anytime soon.

Keywords Central bank · CBDC · Central bank digital currency · Wholesale CBDC · Retail CBDC · Financial stability · Financial crisis · Financial inclusion · Payments

1 Introduction

The evolution of digital currencies dates back to many centuries ago. It began with the use of commodities as money such as copper, bronze, metals, and gold (Kregel 2021). Later, commodity money was abandoned and replaced with fiat money which is paper money that is backed by the government (Gross and Siebenbrunner 2019). Paper money became known as fiat currency or physical currency. The emergence

P. K. Ozili (✉)
Central Bank of Nigeria, Abuja, Nigeria
e-mail: pkozili@cbn.gov.ng

H. Baber and M. Fanea-Ivanovici (eds.), *Sustainable Financing—A Contemporary Guide for Green Finance, Crowdfunding and Digital Currencies*, World Sustainability Series, https://doi.org/10.1007/978-3-031-80969-9_14

of digital technology and its wide applications in the financial services industry in the last two decades led to the digital transformation of money. Private actors began to develop digital forms of money and money alternatives, and it led to the rise of bank money and private digital currencies also known as cryptocurrencies (Khan et al. 2020). A private digital currency refers to currency in digital form in which transactions are verified and records are maintained by a decentralized system using cryptography and controlled by private entities (Khan et al. 2020). The proliferation of private digital currencies led central banks to develop interest in innovating the nature of central bank money in the digital age (Boar and Wehrli 2021). As a result, many central banks are already conducting research into the possibility of issuing a central bank digital currency (CBDC) which is fiat money in digital form that is issued by a central bank (Boar and Wehrli 2021; Ozili 2024).

The purpose of this study is to bring together a collection of informed academic and non-academic perspectives on what a CBDC can achieve and its potential pitfalls. The need for this study arises because recent arguments in favour of, or against, CBDC have moved from research-based debate to a more informed intellectual and emotional debate, and this observation has not been documented in the finance and economics literature. My goal in this study is to present the viewpoints or perspectives of other people as they are, in the form of quotes, and allow the reader to reflect deeply on them, and form an informed opinion about CBDC based on a collection of several views that exist in the public space.

This study is significant because it provides a collection of informed perspectives on some of the recent thinking and research into central bank digital currencies from a diverse variety of people who are truly interested in what a CBDC can achieve and its potential effect on people. This study is valuable to pro-CBDC researchers and central banks who are conducting research into CBDC and are keen to develop a CBDC in the future. This study is also valuable to anti-CBDC individuals and group who think that central banks should not issue a CBDC whether now or in the future.

Many of the perspectives offered through quotes in this article will offer fresh insights into CBDC, and it will reveal unpopular viewpoints that may be helpful in advancing the academic literature on CBDC. Also, expressing the feelings of the general public, through the quote of their representatives, will help the reader to see CBDC in a different light and stimulate the reader to think deeply and more carefully about CBDC. It is hoped that this perspective article will make the reader to ponder deeply on each viewpoint or quote and make the reader to form an objective opinion on what a CBDC can achieve and its potential pitfalls. It is also hoped that this article will provide the reader with new perspectives on other technical aspects of CBDC.

The article is organised in the following way. Section 1 presents the introduction of the study. Section 2 presents the background into central bank digital currency. Section 3 presents the methodology used for the study. Section 4 presents the 100 quotes about CBDC. Section 5 presents the conclusion of the study.

2 Background

In the last five years, many central banks have shown great interest in the possibility of issuing a CBDC due to its potential to give citizens access to central bank reserve money, improve the efficiency of payments, increase financial inclusion, facilitate cross border remittances, and enhance the conduct of monetary policy. Notably, three countries have already issued a CBDC. They include Nigeria which has the eNaira CBDC, Jamaica which has the JAM-DEX CBDC, and the Bahamas which has the Sand Dollar CBDC. These countries felt an urgent need to issue a CBDC and use it as a tool to address their financial exclusion problems and to improve the efficiency of payments. Other central banks, such as the central bank of Brazil, Thailand, the Philippines, Ghana, India, South Korea, Japan, Hong Kong, Singapore, China, the Euro Area, the UK, Russia, Israel and Switzerland, are already conducting CBDC pilot tests to determine the best use case of a CBDC in their countries. Meanwhile, some central banks, such as the U.S. Federal Reserve Bank, have indicated that they will not issue a CBDC in the near future due to lingering issues relating to privacy, trust, the potential for government surveillance as well as other political and legal issues. Despite these issues, a recent Bank of International Settlement (BIS) survey of central banks show that over 93 percent of central banks are involved in some form of CBDC project or research (Kosse and Mattei 2023). This indicates that there is growing certainty that many central banks will issue CBDC in the short to medium term. This positive development, which indicates progress for CBDC, has led to mixed reactions from cryptocurrency advocates, CBDC antagonists and CBDC supporters. I document these mixed reactions in Sect. 4 through quotes. But, before delving into Sect. 4, I would like to explain the methodology used to document these reactions in the form of quotes in the next section.

3 Methodology

The quotes presented in this study were drawn manually from a wide variety of public sources including: media reports; comments on X (formerly known as twitter); expert interview of government officials and private sector experts by media houses; speech of government officials presented at official and unofficial events; publications available on central bank websites; open letters from the general public and associations regarding CBDC; comments of legislators about CBDC during legislative hearings; comments of influential people about CBDC; and published academic and policy research reports about CBDC. The quotes were designed using a similar approach adopted in Ozili (2020) while the quotes obtained from academic research were derived only from the abstract of the research publication, which eliminates the need to provide page numbers for each of the quotes. Regarding the sample period, the quotes were drawn from articles that were published between 2017 to 2024.

Finally, the 100 quotes were carefully selected for information purposes only and all extremely politically sensitive quotes were intentionally excluded.

4 The Quotes About Central Bank Digital Currencies

This section presents some quotes showing optimism about CBDC. It also presents some quotes showing pessimism about CBDC. It further presents some realistic perspectives about CBDC as well as some quotes from academic and policy research. The section also presents some quotes linking cryptocurrency to CBDC.

4.1 Quotes Showing Optimism About CBDC

1. "Central bank digital currencies will become the basis of payments [in the future]."
   – **Jeffrey Sachs**, American Economist, University of Columbia.[1]
2. "Like existing forms of commercial bank money and nonbank money, a CBDC would enable the general public to make digital payments. As a liability of the Federal Reserve, however, a CBDC would not require mechanisms like deposit insurance to maintain public confidence, nor would a CBDC depend on backing by an underlying asset pool to maintain its value. A CBDC would be the safest digital asset available to the general public, with no associated credit or liquidity risk."
   – Statement from a US Federal Reserve Bank Research Report.[2]
3. "A digital euro [CBDC], existing alongside cash, would future-proof our currency. It would be safe, easy to use and free of charge."
   – **Christine Lagarde,** Governor European Central Bank.
4. "[CBDC is] a step forward in democratizing access for individuals and entrepreneurs to the benefit of the digitalization of the economy. Drex [CBDC] will bring more speed, practicality, and lower cost to various contractual and financial transactions we make today."
   – **Maurício Moura**, Deputy Governor, Central Bank of Brazil, August 2023.
5. "There are some key areas where innovation may continue. One that stands out is central bank digital currencies (CBDCs). As the name implies, these would be cryptocurrencies issued directly by central banks. One of their key advantages is that they could act as a form of electronic cash. Some of the more revolutionary possibilities lie in using CBDCs as a foundation on which to build

[1] https://news.bitcoin.com/american-economist-jeffrey-sachs-heralds-end-of-dollar-hegemony-central-bank-digital-currencies-will-become-the-basis-of-payments/.

[2] https://www.federalreserve.gov/publications/money-and-payments-discussion-paper.htm.

a more efficient and interoperable digital domestic and global financial system and we may even see uses no one has even thought of yet appear."

– **Jarek Kowcza,** Senior Economist at Westpac Group.[3]

6. "CBDCs are the catalyst for improving cross-border payments by enabling the build-up of a new international monetary system."

   – **Denis Beau,** Deputy Governor of the Banque de France.
7. "With this [CBDC] pilot project, we are now, for the first time, making it possible to securely and efficiently settle transactions with tokenised assets on a regulated and productive DLT platform using real wholesale CBDC."

   – **Thomas Jordan,** Chairman, Swiss National Bank.[4]
8. "CBDC can operate in places where there is no internet; offline payments will become much easier because from my CBDC wallet I can pay money to your wallet…The greatest advantage of CBDC will be that cross-border payments will become far more efficient, faster and very cost effective. As and when other countries adopt digital currency, I think, international payments systems will really gain in efficiency."

   – **Governor Shaktikanta Das,** Reserve Bank of India. *Responding to comments about India eRupee CBDC during an interaction session at the World Economic Forum Annual Meetings at Davos. January 2024.*
9. "The eNaira [CBDC] is expected to be accessible to all Nigerians. And would provide more possibilities to bring in the unbanked into the digital economy."

   – **Governor Godwin Emefiele,** Central Bank of Nigeria. *Speech at the Grand Finale of the 2022 eNaira Hackathon in Abuja Nigeria. August 2022.*
10. "One should recognize that the CBDC creates new opportunity for monetary policy. If we all had CBDC accounts instead of cash, in principle, it might be possible to implement negative interest rates simply by shrinking balances in CBDC accounts. It will become a lot easier to undertake helicopter drops of money. If everybody had a CBDC account, you could easily increase the balance in those accounts."

    – **Eswar Prasad,** Professor of Trade Policy and Economics, Cornell University. *In an interview on how CBDCs will affect monetary policy. Excerpts from his book: 'The Future of Money: How the Digital Revolution Is Transforming Currencies and Finance.'*
11. "With the proliferation of crypto-moneys, central banks may still be best positioned to provide a cash-like digital currency to the public, as an alternative to digital and physical cash. The Bank of Canada has taken up the challenge of thinking ahead of how such a Central Bank Digital Currency (CBDC) would look like, should there be a political decision to launch one in the future."

    – **Katrin Tinn,** Assistant Professor, Desautels Faculty of Management, McGill University. 2021.

---

[3] https://www.westpac.com.au/news/money-matters/2023/01/cryptocurrencies-miracle-or-mirage/.

[4] https://www.reuters.com/technology/swiss-national-bank-pilot-wholesale-digital-currency-with-ubs-other-banks-2023-11-02/.

12. "We are at an exciting time of innovation in money and payments, and we want to ensure the UK is ready should a decision to build a digital pound [CBDC] be taken in the future."
    – **Bim Afolami,** Economic Secretary to the UK Treasury.[5]
13. "CBDC is not really about replacing commercial banks, but it's ultimately, I think, a furthering up of opportunities for technology-based business models in digital banking."
    – **Alistair Milne**, Professor of Financial Economics at Loughborough University. *In a discussion on CBDC and the fintech ecosystem.*
14. "Monetary policy could act much more directly through a Digital Euro [CBDC], but the decision to ban interest rates on Digital Euros severely restricts these options. Holding limits and the ban on paying interest are instruments that help the Digital Euro fulfil its intended role as a new, public means of payment."
    – **Professor Dirk Niepelt.**[6]
15. "CBDC transactions could also be made anonymous, with access to account-holder information available, when necessary, only to law-enforcement authorities or regulators, as already happens with private banks. Besides, cryptocurrencies such as bitcoin are not actually anonymous, given that individuals and organisations using crypto wallets still leave a digital footprint. And authorities that legitimately want to track criminals and terrorists will soon crack down on attempts to create cryptocurrencies with complete privacy."
    – **Nouriel Roubini**, a professor at NYU's Stern School of Business and CEO of Roubini Macro Associates. *Statement in a 2018 article published in the Guardian titled: Why central bank digital currencies will destroy bitcoin.*
16. "The objective of a CBDC within the framework of the payment system in Peru is to give the unbanked population access to digital payments, so it is important to know their characteristics to prepare an implementation strategy."
    – **A Central Reserve Bank of Peru (BCRP) Report**.
17. "A CBDC can be complementary to the set of existing payment instruments, including banknotes and coins issued by the BCRP, leading to a more efficient, inclusive, and safer payment system."
    – **A Central Reserve Bank of Peru (BCRP) Report**.
18. "By building a robust legal framework [for CBDCs], we can all ensure that CBDCs will flourish."
    – **Agustín Carstens**, General manager of the Bank for International Settlements (BIS).
19. "A digital pound could bring a number of opportunities from potentially enhancing and improving our existing payment systems; providing an alternative to other forms of payment; and potentially spur further innovation across the sector."

[5] https://www.bankofengland.co.uk/news/2024/january/boe-hmt-respond-to-digital-pound-consultation.

[6] https://www.euractiv.com/section/economy-jobs/news/digital-euro-as-currently-planned-is-bound-to-fail-warns-economist/.

– **Lisa Cameron**, Chair of the crypto and digital assets All Party Parliamentary Group. *In a statement to CoinDesk.*

Author's evaluation: The quotes by CBDC optimists indicate that CBDC can transform payments, complement cash, facilitate cross-border payments, lead to efficient offline and online payments, enable financial inclusion for the unbanked, and enhance monetary policy transmission.

4.2 Quotes Showing Pessimism About CBDC

20. "To protect Americans from government tyranny, as your President, I will never allow the creation of a central bank digital currency. Such a currency would give our Federal Government absolute control over your money."
    – **Donald Trump, Former US President.** *In his re-election campaign speech on the election campaign ground in New Hampshire. January 2024.*[7]
21. "The Federal Reserve's exploration into central bank digital currency raises serious questions regarding the continued development of the digital economy, consumer privacy, and the eventual transition to a cashless system of payments. One of the most significant features that draw people to digital assets is decentralization, and there is no central authority that manages the supply and value of most digital assets. The United States must not follow countries like China down the path of digital authoritarianism but instead preserve a payment system that promotes consumer privacy and security."
    – **Adam Brandon,** President and CEO of FreedomWorks. *Response to a proposed bill seeking to ban the Federal Reserve from establishing a CBDC.*[8]
22. "There is a 'don't take my cheese' opposition coming largely from the banks who view the CBDC as a potential disrupter of their very profitable payment systems. It's not a coincidence that pretty much every single bank and every single bank association has been in my office."
    – **Rep. Jim Himes (D-Conn.)**, member of the House Financial Services. *Comment made during a debate about a Fed Digital Dollar. August 2022.*[9]
23. "[A CBDC could] end up becoming the equivalent of the FAA[10] flying planes and building jet engines as opposed to designating competition rules-based safe conduct in the skies."

---

[7] https://twitter.com/BitcoinMagazine/status/1747819823159972003?ref_src=twsrc%5Etfw%7Ctwcamp%5Etweetembed%7Ctwterm%5E1747819823159972003%7Ctwgr%5E350699a00d7479863d6a8c32e717eda728b9d7b5%7Ctwcon%5Es1_amp;ref_url=https%3A%2F%2Fthefintechtimes.com%2Findustry-divided-on-trump-cbdc-views-and-the-impact-no-development-would-have-on-us-crypto-hub%2F.

[8] https://www.cruz.senate.gov/newsroom/press-releases/sen-cruz-introduces-legislation-to-prohibit-the-fed-from-establishing-a-central-bank-digital-currency.

[9] https://www.politico.com/news/2022/08/22/crypto-fed-digital-dollar-00052515.

[10] FAA means the Federal Aviation Administration.

– **Dante Disparte**, Chief Strategy Officer and Head of Global Policy at Circle. *In a statement made to the US House Financial Services Committee.*

24. "A CBDC would remain a liability of the Federal Reserve—remain on the balance sheet of the federal reserve —regardless of how it's distributed because that liability stays on the Federal Reserve's balance sheet, it's effectively a competitor of retail bank deposits and limits the ability of that money to be lent back out into the economy."

    – **Brooke Ybarra**, Head of the American Bankers Association's office of innovation.
25. "CBDCs are a type of government-run cryptocurrency that let the government monitor and control every single transaction you make, no matter how small. That means every single cent can be tracked, traced, taxed, and stopped or even forced without your consent. CBDCs allow bureaucrats to quickly punish you for spreading wrongthink and misinformation. You can be fined, or your bank account can be frozen."

    – **EJ Antoni & Peter St Onge**. *In a commentary on The Heritage Foundation.*[11]
26. "I have yet to see a compelling argument that a U.S. CBDC could solve any of these problems more effectively or efficiently than alternatives, or with fewer downside risks for consumers and for the economy… The potential benefits of a U.S. CBDC remain unclear."

    – **Fed Governor Michelle Bowman**. *Opinion on CBDC in a discourse at Harvard Law School. October 2023.*
27. "A central bank digital currency is government-controlled programmable money that, if not designed to emulate cash, could allow the federal government to surveil and restrict Americans' transactions. This is not just alarming—it's downright un-American."

    – **Tom Emmer, Republican Majority Whip, Minnesota**. *Comments made when introducing the CBDC Anti-Surveillance State Act to the US House of Representatives for consideration. September 2023.*
28. "In response to the popularity of this innovation [cryptocurrency], governments are pursuing the exact opposite: more centralisation, surveillance and control. CBDCs are being developed precisely because they provide governments with increased control and power. This kind of threat to individual rights will naturally drive people toward private solutions, while governments are sure to work hard to thwart such alternatives since they undermine the increased government control and power CBDCs create."

    – **Peter Goettler**, President of the Cato institute.[12]
29. "The Bank of Canada is doing everything necessary to eventually launch a digital currency, even if they still deny wanting to. In their public consultation,

[11] https://www.heritage.org/monetary-policy/commentary/pushers-central-bank-digital-currencies-are-the-most-terrifying-villains.

[12] https://dailyhodl.com/2023/06/28/global-cbdc-rollouts-will-give-governments-massive-control-and-power-warns-cato-institute-president/.

they do mention privacy concerns, but they don't ask a single question about how it would make it easier for them to manipulate monetary policy and control how we spend our money."

– **PPC Leader Maxime Bernier**, speaking to Rebel News.

30. "Governments can program central bank digital currencies (CBDCs) with expiry dates and to restrict undesirable purchases."

    – **Eswar Prasad**, A warning by Cornell University Professor.[13]

31. "Certainly, the way the Bank of England is going, and the Euro system as well, these transactions are not going to be completely anonymous and private because the police and law courts will have the opportunity to jump in if they need to."

    – **Alistair Milne**, Professor of Financial Economics at Loughborough University. In a discussion on CBDC and the fintech ecosystem.

32. "The standard case for a CBDC rests on the mistaken idea that we need new digital money for our new digital world. Much of our money is already digital though; commercial bank deposits and transfers are recorded on computers, not paper ledgers. What we need is not a new digital money, of which there are many already, but instead a new digital cash. Digital cash would not merely facilitate electronic transactions; it would also preserve financial privacy. Even if intermediated, CBDCs are unlikely to fit the bill."

    – **Andrew M. Bailey** (Associate professor of humanities at Yale-NUS College) and William J. Luther (associate professor of economics at Florida Atlantic University). *In a commentary published in a Barron's article.*[14]

33. "The main problem with CBDCs is that they would disrupt the current fractional-reserve system through which commercial banks create money by lending out more than they hold in liquid deposits. Banks need deposits in order to make loans and investment decisions. If all private bank deposits were to be moved into CBDCs, then traditional banks would need to become "loanable funds intermediaries," borrowing long-term funds to finance long-term loans such as mortgages."

    – **Nouriel Roubini**, Professor at NYU's Stern School of Business.[15]

34. "The administration of a CBDC will distract from the Federal Reserve's dual mandate of achieving both stable prices and maximum sustainable employment."

    – **The National Association of Federally-Insured Credit Unions (NAFCU).** *In a public letter to the United States House of Representatives. June 7 2023.*[16]

---

[13] https://crypto.news/cbdcs-can-restrict-citizens-spending-economist-warns/.

[14] https://www.barrons.com/articles/the-U-S-needs-digital-cash-not-just-digital-currency-51646431864.

[15] https://www.theguardian.com/business/2018/nov/19/why-central-bank-digital-currencies-will-destroy-bitcoin.

[16] https://www.nafcu.org/system/files/files/6-7-23%20Digital%20Dollar%20Pilot%20Prevention%20Act%20Support%20Letter.pdf.

35. "CBDCs would threaten the liberties of law-abiding Americans and are being used by authoritarian countries right now to crack down on dissent. That's why closing this pilot program loophole is so important—to prevent the Federal Reserve from bypassing the will of Congress."
    – **Alex Mooney (R-WV)**, Member of the United States House of Representatives.
36. "Current research overwhelmingly undermines the purported benefits of a CBDC and instead indicates that a CBDC would seriously disrupt the financial system, significantly harming consumers and businesses."
    – **Greg Baer**, Bank Policy Institute.
37. "The purported benefits of a CBDC are uncertain and unlikely to be realized, while the costs are real and acute. [CBDC would] rewire our banking and financial system. And even if CBDC accounts were capped at $5,000 per user, they would drain in $720 billion in deposits, or 71% of bank funding. Based on this analysis, we do not see a compelling case for a CBDC in the United States today. Losing this critical funding source would undermine the economics of the banking business model, severely restricting credit availability."
    – **American Bankers Association (ABA)**. The ABA Comments on Federal Reserve Discussion Paper Money and Payments: The US Dollar in the Age of Digital Transformation.

Author's evaluation: The anti-CBDC quotes above suggests that the issuance of a CBDC might distract central banks from their core mandate or it can lead to bank disintermediation. Issuing a CBDC may also lead to government tyranny and a violation of consumer privacy. There is also the perception that there is no clear use case for a CBDC at the moment, and that a CBDC will make central banks be in competition with commercial banks.

4.3 Quotes Showing the Reality of CBDC

38. "Central bank digital currency is changing the very nature of the international monetary system."
    – **Tobias Adrian**, financial counsellor at the International Monetary Fund.[17]
39. "If you're in the financial services sector, this is a potentially worrying time because the arrival of a central bank digital currency (CBDC) in your country will make it much harder to compete in terms of digital payments."
    – **Eswar Prasad**, Professor at Cornell University and Research Associate, National Bureau of Economic Research.
40. "In many countries, privacy concerns are a potential deal breaker when it comes to CBDC legislation and adoption. So, it's vital that policymakers get the mix right."

[17] https://www.centralbanking.com/fintech/cbdc/7958484/cbdcs-will-change-international-monetary-system-imf-panel.

– **Kristalina Georgieva**, *IMF Managing Director. Excerpts from a Speech delivered at the Future of Money: Gearing up for Central Bank Digital Currency event at the Atlantic Council on* 9th *February 2022.*

41. "We are not completely adverse to a UK Central Bank Digital Currency (CBDC) scheme, however, we don't believe a CBDC should be "programmable". This would mean that the issuer of the digital currency would have control over how an individual is permitted to use that currency. We believe that is not right and leave the system open to abuse. We ask the Government to assure that it will prevent any implementation of a "programmable" digital currency."

    – **UK HM Treasury**'s official response to a public petition against introducing a CBDC in the UK.[18]

42. "Many central banks are aware of this—they're aware that a CBDC could limit private sector innovations in digital payments. So, what they're trying to do is provide their central bank digital currencies as tokens, on top of which the private sector can innovate in terms of using those tokens more effectively to intermediate payments between customers and businesses or between businesses and businesses."

    – **Eswar Prasad**, Professor at Cornell University and Research Associate, National Bureau of Economic Research.

43. "If a CBDC were to be issued, it would immediately displace cryptocurrencies, which are not scalable, cheap, secure, or actually decentralised. Enthusiasts will argue that cryptocurrencies would remain attractive to those who wish to remain anonymous. But, like private bank deposits today, CBDC transactions could also be made anonymous, with access to account-holder information available, when necessary, only to law-enforcement authorities or regulators, as already happens with private banks."

    – **Nouriel Roubini**, Professor at NYU's Stern School of Business.[19]

44. "I am fully aware that, in its current state, this [CBDC] initiative raises many questions both for society in general and for the financial industry .... In these circumstances, issuing a digital euro is not naturally perceived to be a pressing need."

    – **Pablo Hernández de Cos**, Bank of Spain Governor.

45. "One aspect of a CBDC that concerns many economists—me included—is the effect it can have on other means of payments issued by private financial institutions, such as commercial banks."

    – **Economist Daniel Sanches**.

46. "The real-world use of CBDCs will depend on how governments address citizens' concerns about government surveillance. While CBDCs can potentially increase financial inclusion and efficiency, distrust of government surveillance could undermine their adoption. Striking the right balance between innovation

[18] https://petition.parliament.uk/petitions/624159.

[19] https://www.theguardian.com/business/2018/nov/19/why-central-bank-digital-currencies-will-destroy-bitcoin.

and privacy is essential to ensure that CBDCs align with citizen's interests rather than becoming a source of discontent and resistance."

– **Nicole Willing**, Technology Journalist. Comment in a post published in Technopedia.[20]

47. "I think we certainly need to understand the technology [behind CBDC], we need to proceed cautiously on building out a rupee CBDC.... There is a need to learn more to understand because in the global arena the rules for CBDCs are now being mapped out and we need to be participants in that rule making so that we do not find ourselves at a disadvantage at some future point."

– **Raghuram Rajan,** Former RBI Governor & Professor at University of Chicago Booth School.[21]

48. "Developing economies seem keener to develop retail CBDC to enhance payment effectiveness and increase financial inclusion, whilst wholesale CBDC have gained traction in more advanced economies where robust and mature interbank systems already exist."

– **Mustafa Syed**. *In an article on LinkedIn titled: Cryptocurrencies and Central Bank Digital Currencies (CBDC) | A Future Outlook.*[22]

49. "It may not solve a problem today, but maybe it solves a problem the day after tomorrow."

– **Dilip Rao**, a former Ripple executive who's now leading Australia's CBDC pilot.[23]

50. "The open infrastructure of the proposed digital pound could enable programmability or automatic transfers with the help of smart contracts should the government allow it."

– **Kene Ezeji-Okoye**, Co-founder at U.K.-based blockchain infrastructure provider Millicent.

51. "If the purpose [of CBDC] is to ensure uniformity of money and protect privacy in a world of digital assets, then it will need to be able to work with the digital assets of the future, and it will need to be just as easy to use."

– **Varun Paul**, Director for CBDC at Fireblocks. *In a statement to CoinDesk.*

52. "Though not in the Bank of England's near-term ambitions, a digital pound [CBDC] could one day unlock real-time and low-cost cross-border transactions directly from their users' digital wallets, eliminating the need for intermediaries and reducing transaction fees."

– **Adam Jackson**, Policy Advisor at Innovate Finance. *In a statement to CoinDesk.*[24]

---

[20] https://www.techopedia.com/will-citizens-distrust-of-government-surveillance-stop-cbdcs.

[21] https://economictimes.indiatimes.com/markets/expert-view/at-present-cryptos-have-little-value-other-than-as-speculative-device-raghuram-rajan/articleshow/96660886.cms?utm_source=contentofinterest&utm_medium=text&utm_campaign=cppst

[22] https://www.linkedin.com/pulse/cryptocurrencies-central-bank-digital-currencies-cbdc-mustafa-syed.

[23] https://cointelegraph.com/news/australia-cbdc-solves-problems-tomorrow-dilip-rao.

[24] https://www.coindesk.com/policy/2023/06/22/digital-pound-should-be-interoperable-with-crypto-uk-lobbyists-say/.

53. "I think that the model that the Bank of England proposes respects the need for privacy and respects the fact that people don't want the government and the central bank to be able to see their transactions."
    – **Jannah Patchay**, Executive Director of the Digital Pound Foundation.
54. "Some members [of The Payments Association] are not comfortable with the central bank being able to view even anonymized transactions/wallet balances. Those that are skeptical of the central bank's and the government's motives might argue that advances in data analysis will allow such data to be interpreted and abused, perhaps through [a] combination with other data sources, including geolocation or tax and benefits data."
    – **Riccardo Tordera-Ricchi**, Head of Policy at The Payments Association. *In a statement to CoinDesk.*[25]

Author's evaluation: The quotes above show that CBDC realists admit that several issues need to be addressed before issuing a CBDC, particularly issues relating to competition with existing payment alternatives, privacy protection, the tyranny of programmable money, preserving the anonymity of money, striking an ethical balance between innovation and privacy, preventing the government from seeing people's transactions in real time and concerns that CBDC may not be a pressing need right now.

4.4 Quotes Linking Cryptocurrency to CBDC

55. "In my view, this [CBDC] development raises an existential problem for cryptocurrencies such as bitcoin: what will be the raison d'etre for cryptocurrencies once CBDCs become ubiquitous? Why would someone prefer to let Facebook's Mark Zuckerberg look after their money rather than Fed Chair Jerome Powell or ECB President Christine Lagarde? And who would prefer to hold their money in an unsecured exchange like Mt. Gox, an entity that was hacked in 2014 while losing $460million in the process? Moreover, why would they choose bitcoin if government-backed alternatives can do it just as quickly, only with a zero bid/offer spread and military-grade security to negate lingering concerns regarding hackable wallets? The answer to this question is that people are likely to go for the easier solution, which is CBDCs."
    – **Marshall Gittler**, Head of Investment Research, BDSwiss. *Statement in an article published on World Finance.*[26]
56. "Insofar as CBDCs would crowd out worthless cryptocurrencies, they should be welcomed. Moreover, by transferring payments from private to central banks, a CBDC-based system would be a boon for financial inclusion. Millions of

[25] https://www.coindesk.com/policy/2023/06/22/digital-pound-should-be-interoperable-with-crypto-uk-lobbyists-say/.

[26] https://www.worldfinance.com/markets/could-central-bank-digital-currencies-spell-the-end-for-cryptocurrency.

unbanked people would have access to a near-free, efficient payment system through their cell phones."

– **Nouriel Roubini**, a professor at NYU's Stern School of Business. In a statement in a 2018 article published in the Guardian titled: Why central bank digital currencies will destroy bitcoin.

57. "If CBDCs are eventually introduced as a counter to cryptocurrencies, it will only be a matter of time until people realise that cryptocurrencies have no real futures as trustworthy currencies, and therefore, are likely to have limited value as assets. But yes, I do wish I had bought a wallet full of bitcoin at 50 cents in 2011."

    – **Marshall Gittler**, Head of Investment Research, BDSwiss. *Statement in an article published on World Finance.*
58. "If [the idea of] a CBDC is moved forward, conventional crypto enthusiasts can rest easy knowing that other types of crypto are still going to be relevant, at least in principle. It's also possible that CBDC can coexist peacefully alongside various other cryptocurrencies."

    – **Deanna Ritchie**. A statement in a post published in Nasdaq.
59. "Crypto has some advantages and some disadvantages, but where are we headed? Many regulators around the world have warned against trading in Bitcoin whereas some countries have already legalized it."

    – **Mustafa Syed**. In an article on LinkedIn titled: Cryptocurrencies and Central Bank Digital Currencies (CBDC) | A Future Outlook.
60. "It's hard to say exactly how CBDC introduction could affect cryptocurrency, but there are a few facts we know. We know Americans overwhelmingly reject this idea. We know there are no immediate plans for the introduction of a CBDC. And we know that the central bank and federal government will have to tread cautiously if they want people to accept this type of digital currency."

    – **Deanna Ritchie**. A statement in a post published in Nasdaq.

Author's evaluation: The quotes above show that there are concerns that the issuance of a CBDC may threaten the existence of cryptocurrencies, and that cryptocurrencies and CBDC may co-exist but not for very long.

4.5 Quotes Showing What the Future Holds for CBDC

61. "The design of the remuneration schedule for CBDC will, in my view, be one of the most significant decisions made by central banks in the next half-century."

    – **Andy Haldane**, the Chief Economist, Bank of England, 2021.[27]
62. "These are still early days for CBDCs and we don't quite know how far and how fast they will go. What we know is that central banks *are* building capacity to harness new technologies—to be ready for what may lie ahead."

[27] https://www.ledgerinsights.com/bank-of-england-economist-cbdc-good-for-negative-interest-rates-narrow-banking/.

– **Kristalina Georgieva**, IMF Managing Director. Excerpts from a Speech delivered at the Future of Money: Gearing up for Central Bank Digital Currency event at the Atlantic Council on 9th February 2022.

63. "Whether or not we will issue a digital shekel [CBDC] is still an open question, as it is in most if not all other advanced economies."

    – **Bank of Israel Governor Amir Yaron**. Speaking at a conference on digital currencies, September 2023.

64. "Country-wide launch will depend on the success, learning and fine tuning of CBDC itself. So, that is the distance which we have to cover. But we have no target date for implementation."

    – **Governor Shaktikanta Das**. Response to comments about India CBDC in an interaction at the World Economic Forum Annual Meetings at Davos.

65. "Trust in all forms of money is an absolute necessity. We know that the decision on whether or not to introduce a digital pound in the UK will be a major one for the future of money."

    – **Shara Breeden**, Deputy Governor for financial stability at the Bank of England.[28]

66. "Maintaining an appropriate level of privacy will be crucial to ensuring public acceptance of retail CBDCs."

    – **Agustin Carstens,** General Manager, Bank of International Settlements (BIS).

67. "We will always ensure people's privacy is paramount in any design [of a CBDC], and any rollout would be alongside, not instead of, traditional cash."

    – **Bim Afolami**, Economic Secretary to the UK Treasury.[29]

68. "The success of a CBDC, if and when issued, will depend on sufficient trust. And, in turn, any successful CBDC should continue to build trust in central banks."

    – **Kristalina Georgieva**, IMF Managing Director. Excerpts from a Speech delivered at the Future of Money: Gearing up for Central Bank Digital Currency event at the Atlantic Council on 9th February 2022.

Author's evaluation: The quotes above regarding the future of CBDC shows that many central banks are still researching CBDC and building capacity for CBDC with the intention of issuing a CBDC in the future. For these central banks, there is no rush to issue a CBDC soon. Other central banks do not intend to issue a CBDC now because they have not resolved issues relating to privacy and trust.

---

[28] https://www.bankofengland.co.uk/news/2024/january/boe-hmt-respond-to-digital-pound-consultation.

[29] https://www.bankofengland.co.uk/news/2024/january/boe-hmt-respond-to-digital-pound-consultation.

4.6 *Quotes from Existing Academic and Policy CBDC Research*

4.6.1 Motivations for CBDC

69. "In order to preserve the effectiveness of monetary policy in a world increasingly flooded by private digital currencies, central banks (CBs) will eventually have to issue their own digital currencies."
    – **Cukierman** (2020). In Abstract, published in *The Manchester School Journal.*
70. "Work on retail CBDCs is more advanced where the informal economy is larger. Many central banks are considering architectures in which a CBDC is a direct cash-like claim on the central bank, but the private sector handles all retail services."
    – **Auer, Cornelli and Frost** (2023). In Abstract, published in *International Journal of Central Banking.*
71. "Countries with more power distance, masculinity and long-term orientation cultures are more eager to adopt a CBDC. By contrast, countries with a cultural value leaning towards uncertainty avoidance are less involved in the adoption of a CBDC."
    – **Luu et al.** (2023a). In Abstract, published in *Applied Economics Letters Journal.*
72. "Perceived usefulness, perceived security, and perceived ease of use, as well as awareness all positively and directly influence individuals' behavioural intentions to use CBDC."
    – **Liu et al.** (2022). In Abstract, published in *Technology Analysis & Strategic Management Journal.*

4.6.2 Benefits of CBDC

73. "The issuance of CBDC is a double-edged sword: the benefit of replacing physical cash for a more efficient payment system comes at the cost of disrupting the deposit base for commercial banks."
    – **Bian, Ji and Wang** (2021). In Abstract, published in *Finance Research Letters Journal.*
74. "CBDC could make monetary policy more efficient, and it could transform the international monetary and payments systems."
    – **Bordo** (2021). Abstract, published in *National Bureau of Economic Research.*
75. "Traders do not view CBDCs as a threat to cryptocurrencies. Rather, more positive stances on CBDCs appear to be interpreted as favorable signals for other forms of digital currencies as well."
    – **Scharnowski** (2022). In Abstract, published in *Finance Research Letters Journal.*

76. "Underdeveloped economies are more engaged in issuing CBDC. Besides, better regulations, FDI inflow, young populations, and more urban societies would increase the probability of CBDC issuance."
    – **Alfar et al.** (2023). In Abstract, published in *Research in International Business and Finance Journal.*
77. "CBDCs promise to realize a broad range of new capabilities, including direct government disbursements to citizens, frictionless consumer payment and money-transfer systems, and a range of new financial instruments and monetary policy levers. A CBDC will also in some way need to address an innate tension between privacy and transparency, protecting user data from abuse while selectively permitting data mining for end-user services, policymakers, and law enforcement investigations and interventions."
    – **Allen et al.** (2020). In Abstract, published in *the National Bureau of Economic Research.*

4.6.3 Cash Usage

78. "CBDC may not be widely accepted by end users in the presence of a sizable informal economy in which cash is the primary method of payment. CBDC can decrease informality, but this effect becomes weaker in countries with larger informal economies."
    – **Oh and Zhang** (2022). In Abstract, published in *Economic inquiry Journal.*
79. "A cash-like CBDC is more effective than a deposit-like CBDC in promoting consumption and welfare. Interestingly, a cash-like CBDC can also crowd in banking, even in the absence of bank market power. In a calibrated model, at the maximum, a cash-like CBDC can increase bank intermediation by 10.2% and welfare by 0.059%, and it can capture up to 23.3% of the payment market."
    – **Chiu and Davoodalhosseini** (2023). In Abstract, published in *Management Science Journal.*
80. "A CBDC can be designed with attributes similar to cash or deposits and can be interest bearing: a CBDC that closely competes with deposits depresses bank credit and output, while a cash-like CBDC may lead to the disappearance of cash. Then, the optimal CBDC design trades off bank intermediation against the social value of maintaining diverse payment instruments."
    – **Agur, Ari and Dell'Ariccia** (2022). In Abstract, published in *Journal of Monetary Economics.*
81. "Having both cash and a CBDC available [for use in the economy] may result in lower welfare than in the cases where only cash or only a CBDC is available. If the cost of using a CBDC relative to cash is around 0.25% of the transaction value, introducing a CBDC can lead to an increase of 0.12–0.21% consumption for the United States and 0.04–0.07% for Canada."
    – **Davoodalhosseini** (2022). Abstract, published in *Journal of Economic Dynamics and Control.*

4.6.4 Financial Stability

82. "If the introduction of CBDC follows a set of conservative core principles, bank funding is not necessarily reduced, credit and liquidity provision to the private sector need not contract, and the risk of a system-wide run from bank deposits to CBDC is addressed."
– **Kumhof and Noone** (2021). Abstract, published in *Economic Analysis and Policy Journal.*
83. "Banks [will] do less maturity transformation when depositors have access to CBDC, which leaves them less exposed to [bank] runs. Second, monitoring the flow of funds into CBDC allows policymakers to identify and resolve weak banks sooner, which also decreases depositors' incentive to run [on a bank]."
– **Keister and Monnet** (2022). In Abstract, published in *Journal of Economic Dynamics and Control.*
84. "However, once the central bank can lend all the deposits in CBDC account to banks, an increase in the quantity of CBDC which does not require reserve holdings can enhance financial stability by increasing credit supply and lowering nominal interest rate."
– **Kim and Kwon** (2023). In Abstract, published in *Journal of Money, Credit and Banking.*
85. "The introduction of deposits in CBDC account decreases credit supply by banks, raising the nominal interest rate and lowering a bank's reserve-deposit ratio. This increases the likelihood of bank panic in which banks exhaust cash reserves."
– **Kim and Kwon** (2023). In Abstract, published in *Journal of Money, Credit and Banking.*
86. "If the central bank aims to have a positive net worth and the absence of bank runs, a high demand for a CBDC is a necessary condition to achieve both objectives. If this is not the case, financial stability will be endangered."
– **Tercero-Lucas** (2023). In Abstract, published in *Journal of Financial Stability.*
87. "Higher CBDC remuneration raises bank fragility by increasing consumers' withdrawal incentives. On the other hand, it also induces banks to offer more attractive deposit contracts in order to retain funding, thereby reducing fragility. This results in a U-shaped relationship between bank fragility and CBDC remuneration."
– **Ahnert et al.** (2023). In Abstract, published in the *CEPR Discussion Papers.*
88. "Issuing a CBDC by expanding the central bank's balance sheet can have a durable impact on GDP and crucially, it allows for the activation of a new transmission channel, which depends on the sensitivity of investment and of the interest rate to firms' leverage. But it might reduce the demand for bank deposits, and even more crucially for bank loans, thus creating challenges for financial stability."

– **Temperini, D'Ippoliti and Gobbi** (2024). In Abstract, published in *Structural Change and Economic Dynamics Journal.*

4.6.5 Monetary Policy

89. "Central bank digital currency (CBDC) is seen as a possible next step in the evolution of money, offering a more stable unit of account, a more efficient medium of exchange, and a safer way to store value."

    – **Hoang et al.** (2023). In Abstract, published in *Research in International Business and Finance Journal.*

90. "If cash is important for monetary policy, payment instrument competition, or as an alternative payment instrument in the event of operational problems with privately supplied payment methods, the introduction of CBDC may best be introduced before cash substitutes become so ubiquitous that the viability of CBDC could be in doubt."

    – **Khiaonarong and Humphrey** (2019). In Abstract, published in *Journal of Payments Strategy and Systems.*

91. "CBDC issuance of 30% of GDP, against government bonds, could permanently raise GDP by 3%, due to lower real interest rates, distortionary taxes, and monetary transaction costs. Countercyclical CBDC policy rules, as a second monetary policy tool, could substantially improve the central bank's ability to stabilise the business cycle."

    – **Barrdear and Kumhof** (2022). In Abstract, published in *Journal of Economic Dynamics and Control.*

92. "The central bank can implement negative interest rate policy by directly adjusting the interest rate of digital currency to stimulate consumption, investment, and output and to accelerate macroeconomic recovery."

    – **Xin and Jiang** (2023). In Abstract, published in *Research in International Business and Finance Journal.*

4.6.6 Effect on Financial Institutions

93. "The issuance of a CBDC is a double-edged sword for an economy. Although it provides additional payment and investment options for market participants, it increases competition and weakens financial intermediaries' margins."

    – **Son, Ryu and Webb** (2023). In Abstract, published in *International Review of Financial Analysis.*

94. "The introduction of a central bank digital currency has no detrimental effect on bank lending activity and may, in some circumstances, even serve to promote it."

    – **Andolfatto** (2021). In Abstract, published in *The Economic Journal.*

95. "While a digital currency tends to improve efficiency in exchange, it may also crowd out bank deposits, raise banks' funding costs, and decrease investment."

– **Keister and Sanches** (2023). In Abstract, published in *The Review of Economic Studies.*

96. "CBDC helps to reduce leverage and asset risks. The adoption of CBDC can be associated with expanded lending, increased asset quality and reduced loan loss reserves."

    – **Luu et al.** (2023b). In Abstract, published in *Journal of International Financial Markets, Institutions and Money.*

97. "Central banks wishing to introduce an economy-wide, account-type CBDC should first monitor yields on bank loans and consider policy measures that induce banks to maintain adequate liquidity reserve levels."

    – **Jun and Yeo** (2021). In Abstract, published in *Financial Innovation Journal.*

98. "Depending on its design, a digital euro [CBDC] could improve bank profitability and competitiveness by absorbing large amounts of idle (and expensive) excess reserves without penalizing lending, while incentivizing bank digitalization."

    – **Fegatelli** (2022). In Abstract, published in *Journal of Macroeconomics.*

99. "If banks have market power in the deposit market, a CBDC can enhance competition, raising the deposit rate, expanding intermediation, and increasing output. A CBDC can raise bank lending by 1.57% and output by 0.19% [in the US economy]."

    – **Chiu et al.** (2023). In Abstract, published in *Journal of Political Economy.*

100. "Negative media sentiment about [CBDCs such as] the eNaira, DCash and Sand Dollar [is] caused by the existing negative media sentiment about CBDCs in general."

    Ozili and Alonso (2024). In Abstract, published in *Journal of Central Banking Theory and Practice.*

Author's evaluation: Quotes from academic and policy research offer interesting technical insights into the motivations for issuing a CBDC, the benefits of CBDC, its effect on the use of cash, the financial stability implications, the monetary policy implications, and the effect of CBDC on financial institutions.

5 Conclusion

This study presented several informed views about the introduction of a central bank digital currency (CBDC) in the form of quotes. The significance of this perspective study is that it provides a collection of informed perspectives on the recent thinking and research about central bank digital currencies from people who are most concerned about CBDC. To date, this type of perspective study has not been documented in the central bank digital currency literature. This perspective study is most valuable to central bankers, economists and researchers who want to gain useful insights into CBDC and those who want to assist their country in developing a national CBDC. The starting point of such endeavour is to listen to the perspective of informed promoters and antagonists of CBDC. The insights gained can help them to develop a meaningful central bank digital currency.

My goal in this study was to present the views as they are, in the form of quotes, and allow the reader to reflect deeply on them, and form an informed opinion about CBDC based on a collection of several views that exist in the public space. From the quotes shown in this study, it is easy to see that a CBDC is the digital equivalent of physical cash, and it is issued by the central bank. It is also easy to see that the concept of a CBDC has come to stay and that many central banks want to issue a CBDC in the distant future. It is also easy to see, from the quotes, that despite the efforts of central banks and pro-CBDC enthusiasts academic researchers to publicise the benefits of a CBDC to citizens, strong concerns have been raised about CBDC particularly the potential for government overreach and surveillance, loss of competitive advantage for deposit-taking financial institutions, loss of privacy for citizens, and concerns that CBDC development is an unwholesome distraction for central banks, among others.

Due to these concerns, there is widespread negative media campaign to turn ordinary citizens against the issuance of a CBDC. Such negative media campaign is unfair to central banks because central banks, just like any organisation, have a right to introduce innovation. But the ongoing negative media campaign against CBDC are intended to bring an end to CBDC innovation before it is launched. Central banks should be allowed to issue a CBDC, and let the people decide whether the CBDC will be successful or a failure. If majority of the population refuse to use the issued CBDC, then the central bank must discontinue the CBDC on the merit that people have refused to use it. Similarly, if majority of the population accept to use the issued CBDC, then the central bank should improve the issued CBDC on the merit that people have willingly accepted to use CBDC. I think this is the proper posture or attitude to take towards CBDC. We need to inform people about the benefits and disadvantages of using a CBDC and allow the people to decide whether they will use the CBDC or not. For central banks that are keen on developing a CBDC, they should encourage more debates and deliberations on the best way to develop a CBDC that is non-intrusive, ensures privacy, complement existing alternatives, and prevent government surveillance.

Acknowledgements None.

Competing Interests
The author has no conflicts of interest to declare that are relevant to the content of this chapter. There is no financial or non-financial competing interest associated with this research.

References

Agur I, Ari A, Dell'Ariccia G (2022) Designing central bank digital currencies. J Monet Econ 125:62–79

Ahnert T, Hoffmann P, Leonello A, Porcellacchia D (2023) Central bank digital currency and financial stability (No. 18222). CEPR Discussion Papers

Alfar AJ, Kumpamool C, Nguyen DT, Ahmed R (2023) The determinants of issuing central bank digital currencies. Res Int Bus Financ 64:101884

Allen S, Čapkun S, Eyal I, Fanti G, Ford BA, Grimmelmann J, Juels A, Kostiainen K, Meiklejohn S, Miller A, Prasad E (2020) Design choices for central bank digital currency: policy and technical considerations (No. w27634). National Bureau of Economic Research

Andolfatto D (2021) Assessing the impact of central bank digital currency on private banks. Econ J 131(634):525–540

Auer R, Cornelli G, Frost J (2023) Rise of the central bank digital currencies. Int J Cent Bank 19(4):185–214

Barrdear J, Kumhof M (2022) The macroeconomics of central bank digital currencies. J Econ Dyn Control 142:104148

Bian W, Ji Y, Wang P (2021) The crowding-out effect of central bank digital currencies: a simple and generalizable payment portfolio model. Financ Res Lett 43:102010

Boar C, Wehrli A (2021) Ready, steady, go?-Results of the third BIS survey on central bank digital currency. BIS papers

Bordo MD (2021) Central bank digital currency in historical perspective: Another crossroad in monetary history (No. w29171). National Bureau of Economic Research

Chiu J, Davoodalhosseini SM (2023) Central bank digital currency and banking: macroeconomic benefits of a cash-like design. Manage Sci 69(11):6708–6730

Chiu J, Davoodalhosseini SM, Jiang J, Zhu Y (2023) Bank market power and central bank digital currency: theory and quantitative assessment. J Polit Econ 131(5):1213–1248

Cukierman A (2020) Reflections on welfare and political economy aspects of a central bank digital currency. Manch Sch 88:114–125

Davoodalhosseini SM (2022) Central bank digital currency and monetary policy. J Econ Dyn Control 142:104150

Fegatelli P (2022) A central bank digital currency in a heterogeneous monetary union: managing the effects on the bank lending channel. J Macroecon 71:103392

Gross MM, Siebenbrunner C (2019) In: Money creation in fiat and digital currency systems. International Monetary Fund

Hoang YH, Ngo VM, Vu NB (2023) Central bank digital currency: A systematic literature review using text mining approach. Res Int Business and Finance 101889

Jun J, Yeo E (2021) Central bank digital currency, loan supply, and bank failure risk: a microeconomic approach. Finan Innov 7(1):1–22

Keister T, Monnet C (2022) Central bank digital currency: stability and information. J Econ Dyn Control 142:104501

Keister T, Sanches D (2023) Should central banks issue digital currency? Rev Econ Stud 90(1):404–431

Khan MZ, Ali Y, Sultan HB, Hasan M, Baloch S (2020) Future of currency: a comparison between traditional, digital fiat and cryptocurrency exchange mediums. Int J Blockchains and Cryptocurrencies 1(2):206–224

Khiaonarong T, Humphrey D (2019) Cash use across countries and the demand for central bank digital currency. J Payments Strategy Syst 13(1):32–46

Kim YS, Kwon O (2023) Central bank digital currency, credit supply, and financial stability. J Money, Credit, Bank 55(1):297–321

Kosse A, Mattei I (2023) Making headway-results of the 2022 BIS survey on central bank digital currencies and crypto. BIS Papers, No. 136

Kregel J (2021) The economic problem: from barter to commodity money to electronic money. Levy Economics Institute, Working Papers Series

Kumhof M, Noone C (2021) Central bank digital currencies—Design principles for financial stability. Econ Anal Policy 71:553–572

Liu X, Wang Q, Wu G, Zhang C (2022) Determinants of individuals' intentions to use central bank digital currency: evidence from China. Technol Anal Strategic Managem 1–15
Luu HN, Do DD, Pham T, Ho VX, Dinh QA (2023a) Cultural values and the adoption of central bank digital currency. Appl Econ Lett 30(15):2024–2029
Luu HN, Nguyen CP, Nasir MA (2023b) Implications of central bank digital currency for financial stability: evidence from the global banking sector. J Int Finan Markets Inst Money 89:101864
Oh EY, Zhang S (2022) Informal economy and central bank digital currency. Econ Inq 60(4):1520–1539
Ozili PK, Alonso SLN (2024) Central Bank digital currency adoption challenges, solutions, and a sentiment analysis. J Central Banking Theory and Pract 13(1):133–165
Ozili PK (2020) 100 Quotes from the Global Financial Crisis: Lessons for the Future. In: Uncertainty and challenges in contemporary economic behaviour (pp. 185–194). Emerald Publishing Limited
Ozili PK (2024) Central Bank digital currency and the monetary policy and financial stability implications. In: Digital currency assets and challenges to financial system stability. IGI Global, pp 52–62
Scharnowski S (2022) Central bank speeches and digital currency competition. Financ Res Lett 49:103072
Son J, Ryu D, Webb RI (2023) Central bank digital currency: payment choices and commercial bank profitability. Int Rev Financ Anal 90:102874
Temperini J, D'Ippoliti C, Gobbi L (2024) Is the time ripe for helicopter money? Growth impact and financial stability risks of outright monetary transfers. Struct Chang Econ Dyn 69:24–36
Tercero-Lucas D (2023) Central bank digital currencies and financial stability in a modern monetary system. J Financ Stab 69:101188
Xin B, Jiang K (2023) Central bank digital currency and the effectiveness of negative interest rate policy: a DSGE analysis. Res Int Bus Financ 64:101901

MIX
Papier aus verantwortungsvollen Quellen
Paper from responsible sources
FSC® C105338

If you have any concerns about our products,
you can contact us on
ProductSafety@springernature.com

In case Publisher is established outside the EU,
the EU authorized representative is:
Springer Nature Customer Service Center GmbH
Europaplatz 3, 69115 Heidelberg, Germany

Printed by Libri Plureos GmbH
in Hamburg, Germany